Fabrication and Properties of Functional Coatings Under Extreme Conditions

Fabrication and Properties of Functional Coatings Under Extreme Conditions

Guest Editor

Guoqin Cao

MDPI

Basel • Beijing • Wuhan • Barcelona • Belgrade • Novi Sad • Cluj • Manchester

Guest Editor
Guoqin Cao
School of Materials Science
and Engineering
Zhengzhou University
Zhengzhou
China

Editorial Office
MDPI AG
Grosspeteranlage 5
4052 Basel, Switzerland

This is a reprint of the Special Issue, published open access by the journal *Materials* (ISSN 1996-1944), freely accessible at: https://www.mdpi.com/journal/materials/special_issues/5T692889ZC.

For citation purposes, cite each article independently as indicated on the article page online and as indicated below:

Lastname, A.A.; Lastname, B.B. Article Title. *Journal Name* **Year**, *Volume Number*, Page Range.

ISBN 978-3-7258-7524-5 (Hbk)
ISBN 978-3-7258-7525-2 (PDF)
https://doi.org/10.3390/books978-3-7258-7525-2

Contents

About the Editor

Guoqin Cao

Guoqin Cao is affiliated with the School of Materials Science and Engineering, Zhengzhou University, Zhengzhou 450001, China. His research interests cover a wide range of fields related to functional materials, with a primary focus on TEM characterization, films and coatings, material surface and interface behavior, and oxidation mechanisms. He has long been dedicated to exploring the preparation technologies, performance regulation, and failure mechanisms of coatings under extreme service conditions, such as aerospace, marine, and nuclear power environments. His work is committed to revealing the intrinsic relationship of the composition and structure of coating materials with their service stability, providing important theoretical guidance and technical support for the development and application of advanced functional coating materials.

Preface

This Reprint compiles cutting-edge research papers centered on the fabrication and properties of functional coatings under extreme conditions, drawn from the "Thin Films and Interfaces" Section in *Materials*. In recent decades, the rapid development of high-end industries such as aerospace, marine engineering, and nuclear power has resulted in increasingly stringent requirements for the performance of key materials, which are often exposed to extreme service environments involving intense thermal shock, severe corrosion, and high mechanical stress. As a core technology for enhancing the durability, stability, and functional characteristics of substrate materials, functional coatings have become a research hotspot, attracting extensive attention from both academia and industry. Against this backdrop, this Reprint presents the latest progress in the processing, characterization, and application of thin films and coatings, with a focus on addressing the technical bottlenecks and scientific issues faced by coating materials in extreme environments.

Guoqin Cao
Guest Editor

materials

Editorial

Fabrication and Properties of Functional Coatings Under Extreme Conditions

Chaoyang Chen [1], Lingjie Chen [2,*] and Guoqin Cao [1,*]

1 State Centre for International Cooperation on Designer Low-Carbon & Environmental Materials, School of Materials Science and Engineering, Zhengzhou University, Zhengzhou 450001, China
2 National Key Laboratory of Marine Corrosion and Protection, Luoyang Ship Material Research Institute, Luoyang 471023, China
* Correspondence: cljie725@163.com (L.C.); caoguoqin@zzu.edu.cn (G.C.)

The advancement of aerospace, marine engineering, and advanced energy systems hinges on material performance under extreme service conditions (intense thermal exposure, severe corrosion, high mechanical stress) [1,2]. Conventional materials/coatings fall short here—limitations in thermal stability, corrosion resistance, and mechanical integrity under multi-factor stress create performance gaps and operational risks, even leading to catastrophic equipment damage and economic loss. Thus, developing next-generation functional coatings (with enhanced durability, multifunctionality, adaptive protection) is both a scientific priority and a technological imperative for these sectors [3–5]. Future progress will depend on innovative material architectures and advanced fabrication technologies: key directions include self-healing coatings (based on stimuli-responsive micro/nanocontainers), multifunctional materials (integrating corrosion inhibition, lubricity), and emerging synthesis methods supported by advanced theoretical modeling. These developments are critical for enabling applications at the limits of material capability [6–8].

This Special Issue focuses particularly on service behavior under extreme nuclear, aerospace, and marine environments, alongside the preparation and implementation of emerging materials—including nanocomposites, multilayer architectures, high-entropy alloys, and high-entropy ceramics. These advanced materials have been extensively investigated in recent years due to their exceptional structural stability and environmental adaptability [9–11]. Contributions addressing the characterization of complex oxidation processes, interfacial interactions, and the correlation between composition, structure, and in-service stability are especially welcome.

The ten articles included in this Special Issue collectively establish an integrated research chain spanning "fabrication–characterization–application under extreme conditions". Each contribution addresses specific industrial challenges while deepening fundamental insights. In the context of extreme wear environments, such as those encountered by mining machinery, molds, and cutting tools subjected to high loads and abrasive dust, two studies are particularly noteworthy. Piotrowska et al. [contribution 1] examined the tribological performance of diamond-like carbon (DLC) coatings of the a-C:H type, deposited by plasma-assisted chemical vapor deposition, in comparison with conventional 100Cr6 steel in mine conveyor belt bearings. Their results indicated that the DLC coating reduced the friction coefficient by more than 30%, decreased the wear rate to one-fifth that of 100Cr6 steel, and doubled the mean time between failures, thereby substantially lowering equipment downtime losses. In another study, Hsu et al. [contribution 2] fabricated (Ti, Cr, Cu, Al, Si)N multi-element nitride coatings on AISI H13 tool steel via cathodic

 https://doi.org/10.3390/ma18225231

arc deposition. By optimizing the N_2/Ar flow ratio to 2, they obtained a dense solid-solution structure exhibiting a hardness of 28 GPa, i.e., three times that of the substrate, along with interfacial adhesion exceeding 50 N and a 40% reduction in wear loss during high-temperature cutting operations. Notably, these performance gains were achieved while maintaining low surface roughness (Ra < 0.5 μm), essential for preserving machining precision.

For applications in high-temperature and nuclear environments, such as nuclear accident-tolerant fuel claddings and aero-engine blades, where exceptional oxidation resistance is critical, a collaborative study conducted by Dr. Guoqin Cao [contribution 3] investigated Al-doped TiCrZrNbTa refractory high-entropy alloy (RHEA) coatings. These coatings were deposited onto Zry-4 substrates via magnetron sputtering. Oxidation tests conducted in high-temperature steam at 1000–1100 °C revealed that a composition of 10 at.% Al promoted the formation of a dense composite oxide scale (Al_2O_3–TiO_2–Cr_2O_3), which reduced the oxidation weight gain to just one-tenth of that observed in Al-free coatings, while maintaining interfacial adhesion strength above 45 MPa without evidence of cracking. In contrast, Al concentrations equal to or exceeding 17 at.% resulted in the formation of brittle Al-rich phases, which accelerated both oxidation and spallation. These findings delineate a critical "Al threshold" in the compositional design of high-entropy alloy coatings for extreme service conditions.

Addressing corrosion challenges in marine and acidic environments, where high-salt sprays and industrial acids degrade metallic substrates, three studies in this issue present innovative coating solutions. Wenqiang Li et al. [contribution 4] fabricated $Fe_{60}Co_{10-x}Ni_{15}Cr_{15}Si_x$ coatings on Q235 steel via a combined aerosol powder synthesis and laser cladding route, demonstrating that 4 at.% Si optimally balanced corrosion resistance and cost—achieving a corrosion current density only 1/20 that of bare Q235 steel in 3.5% NaCl solution, with a 15% cost reduction compared to Si-free counterparts. In contrast, higher Si content (8 at.%) compromised performance due to its intrinsic poor corrosion resistance. Aleksandra Kucharczyk-Kotlewska et al. [contribution 5] developed VTMS/PEDOT/PMo12 composite coatings for X20Cr13 and 41Cr4 steels, which reduced the corrosion rate by over 90% in pH 2 acidic medium—attributed to PMo_{12}'s ion barrier functionality and PEDOT's ability to facilitate passive film formation. Meanwhile, Agata Dudek et al. [contribution 6] engineered VTMS/HAp/VTMS/VTMS multilayer coatings on Grade 2 titanium using a sol–gel process, achieving interfacial adhesion of 30 N, a corrosion current density 50 times lower than that of pure titanium in simulated body fluid, and enhanced cytocompatibility via hydroxyapatite (HAp) incorporation—offering an integrated coating strategy for biomedical implants such as artificial joints.

In a study that bridges corrosion protection and novel fabrication methods, Pelucchi et al. [contribution 7] investigated a direct electrochemical route for depositing graphene onto steel substrates via cyclic voltammetry. This innovative approach bypasses the traditional, environmentally taxing graphene oxide (GO) synthesis step, enabling in situ oxidation and reduction in graphene flakes directly on the metal surface. While the coating quality of this direct method currently does not surpass that of conventional GO-based coatings, it presents a significant advancement in process sustainability and simplification, pointing toward greener fabrication routes for protective graphene coatings.

Beyond traditional protection, two articles expand functional coatings to optics and textiles. Daofeng Zhu et al. [contribution 8] synthesized amphiphilic polyacrylate (PMLEA) via solution radical copolymerization, optimizing the LA/ER-10 mass ratio to 1:2 to create a peelable coating that completely removes dust, grease, and fingerprints from quartz glass without residue, while maintaining excellent mechanical properties (2.57 MPa tensile strength, 183% elongation at break) and temperature stability (−20–80 °C) for high-precision optical lenses and spacecraft sensors. Mariia Svyntkivska et al. [contribution 9] coated cotton and PET fabrics with graphene oxide (GO) via padding, using quercetin (a natural flavonoid) for thermal reduction—achieving surface resistances as low as 240 kΩ/sq (PET) and 750 kΩ/sq (cotton), hydrophobicity (163° and 147° water contact angles, respectively), and over 99% antibacterial rates, supporting smart and medical textiles.

Finally, addressing ceramic-metal bonding challenges, which are critical for solid oxide fuel cells and high-temperature sensors, Shu-Wei Chang et al. [contribution 10] deposited 1.5–6 μm Cu-Ti coatings (70.2Cu-29.8Ti at.%) on ZrO_2 and Crofer via vacuum sputtering then used silver-based brazing for reactive air brazing (RAB). A 3 μm coating formed "Ti-O-Zr" chemical bonds with ZrO_2 and solid solutions with Crofer, eliminating interfacial defects and maintaining hermeticity (no leakage) for 28 h at 2 psig (room temperature) and 24 h at 600 °C—exceeding industry standards and advancing fuel cell industrialization.

In conclusion, this Special Issue compiles the latest advances in functional coatings for extreme environments, integrating material innovation, process optimization, and mechanistic studies to address industrial challenges and outline future research directions—such as coatings for multi-factor coupled extreme conditions, atomic-level interface regulation, and coating-substrate synergy. Academically, it establishes quantitative thresholds for multi-component coatings, develops eco-friendly processes, and builds a robust performance prediction framework. Industrially, it meets critical needs across five strategic sectors: energy, manufacturing, marine, medical, and aerospace. As an open-access resource, it promotes global collaboration and technology transfer. Dr. Guoqin Cao, the Guest Editor, wishes to express gratitude to all authors, reviewers, and the *Materials* editorial team. "We aim for this Special Issue to serve as a key academic platform for the extreme-environment coating community, driving interdisciplinary collaboration and innovation to advance materials protection for global high-end equipment." The Second Edition of Fabrication and Properties of Functional Coatings Under Extreme Conditions is on https://www.mdpi.com/journal/materials/special_issues/HV8A3HG7YG (accessed on 16 November 2025).

Author Contributions: C.C. conceived the investigation and wrote the original draft; and G.C. and L.C. were responsible for formal analysis, participated in writing—review and editing. All authors have read and agreed to the published version of the manuscript.

Funding: This work was partly funded by the National Natural Science Foundation of China [52403406], the Natural Science Foundation of Henan [232300421013], Postgraduate Education Reform Project of Henan Province [2023SJGLX030Y] and the Department of Science and Technology of Henan Province [232301420001].

Conflicts of Interest: The authors declare no conflicts of interest.

List of Contributions:

1. Piotrowska, K.; Madej, M.; Milewski, K. Tribological Properties of DLC Coatings in Model-Based and Real-Life Tests. *Materials* **2025**, *18*, 4251.
2. Pelucchi, M.; Alfano, B.; Lama, G.C.; Rosa, R.P.; Cabrini, M. Preliminary Study on Electrochemical Deposition of Graphene on Steel Substrate via In Situ Oxidation Using Cyclic Voltammetry. *Materials* **2025**, *18*, 2440.
3. Guo, M.; Chen, C.; Song, B.; Guo, J.; Hu, J.; Cao, G. Effect of Al on the Oxidation Behavior of TiCrZrNbTa High-Entropy Coatings on Zr Alloy. *Materials* **2025**, *18*, 1997.
4. Li, W.; Lian, J.; Wang, D.; Zhang, S.; Han, C.; Du, Z.; Li, F. Effect of Si Addition on Structure and Corrosion Resistance of FeCoNiCr High-Entropy Alloy Coating. *Materials* **2024**, *18*, 72.
5. Kucharczyk-Kotlewska, A.; Adamczyk, L.; Miecznikowski, K.; Dudek, A. Protective Properties of Silane Composite Coatings Modified with Poly (3, 4-ethylenedioxythiophene) with Heteropolyacid on X20Cr13 and 41Cr4 Steel. *Materials* **2024**, *17*, 6177.

6. Dudek, A.; Kierat, O. Preliminary Aspects Regarding the Anticorrosive Effect of Multi-Layered Silane–Hydroxyapatite Coatings Deposited on Titanium Grade 2 for Medical Applications. *Materials* **2024**, *17*, 6001.
7. Zhu, D.; Huang, H.; Liang, A.; Yang, Y.; He, B.; Ahmed, A.; Li, X.; Ding, F.; Sun, L. Synthesis of Amphiphilic Polyacrylates as Peelable Coatings for Optical Surface Cleaning. *Materials* **2024**, *17*, 4813.
8. Hsu, C.-H.; Chen, H.-W.; Lin, C.-Y.; Hu, S.-H. Effect of N2/Ar ratio on wear behavior of multi-element nitride coatings on AISI H13 tool steel. *Materials* **2024**, *17*, 4748.
9. Chang, S.-W.; Shiue, R.-K.; Huang, L.-W. Zirconia and crofer joint made by reactive air brazing using the silver base paste and Cu-Ti coating layer. *Materials* **2024**, *17*, 3822.
10. Svyntkivska, M.; Makowski, T.; Kregiel, D.; Piorkowska, E. Electrical Conductivity and Antibacterial Activity of Woven Fabrics through Quercetin-Assisted Thermal Reduction of a Graphene Oxide Coating. *Materials* **2023**, *16*, 7184.

References

1. Eswarappa Prameela, S.; Pollock, T.M.; Raabe, D.; Meyers, M.A.; Aitkaliyeva, A.; Chintersingh, K.-L.; Cordero, Z.C.; Graham-Brady, L. Materials for extreme environments. *Nat. Rev. Mater.* **2023**, *8*, 81–88. [CrossRef]
2. Xie, F.; Yang, Y.; Ping, Z.; Gong, X.; Dong, Z.; Xu, W.; Yu, J.; Yao, W.; Dong, Y. Progress and strategies in advanced aerospace materials for extreme environments: A review. *Acta Astronaut.* **2026**, *238*, 184–196. [CrossRef]
3. Summers, W.D.; Begley, M.R.; Zok, F.W. Transition from penetration cracking to spallation in environmental barrier coatings on ceramic composites. *Surf. Coat. Technol.* **2019**, *378*, 125083. [CrossRef]
4. Wan, S.; Cui, X.; Jin, G.; Ma, J.; Yang, Y.; Liu, K.; Li, J.; Wang, S.; Wang, J. Microstructure and properties characterization of laser-cladded CuAl alloy coatings on MgLi alloy. *Surf. Coat. Technol.* **2023**, *460*, 129430. [CrossRef]
5. Hao, S.; Oleksak, R.P.; Doğan, Ö.N.; Gao, M.C. Computational Design of High-Entropy Rare Earth Disilicates as Next-Generation Thermal/Environmental Barrier Coatings. *Acta Mater.* **2023**, *258*, 119225. [CrossRef]
6. Zhang, X.; Gong, X.; Wang, D.; Liu, S.; Zhang, Y. Stimuli-responsive micro/nanocontainers for multifunctional anticorrosive coatings: Design, mechanism, and applications. *Prog. Org. Coat.* **2026**, *210*, 109712. [CrossRef]
7. Li, Z.; Liu, Y.; Pei, X.; Wu, Y.; Zhou, F. Environmentally adaptive multifunctional binary-polymer-brushes coating with aqueous superlubricity and dry self-lubrication. *Chem. Eng. J.* **2025**, *523*, 168445. [CrossRef]
8. He, C.; Pan, D.; Li, X.; Lu, Z.; Chen, K.; Wang, G.; Zhang, Z.; Zhang, H.; Zhang, Y.; Wen, Z. Fresh perspectives and insights into the challenges and opportunities in the emerging high-entropy electrocatalysts. *Coord. Chem. Rev.* **2025**, *531*, 216496. [CrossRef]
9. Wang, Z.; Zhang, Y.; Han, Z.; Wang, Z.; Yang, W.; Li, M.; Wang, A.; Ke, P. Ultra-high resistance to electrochemical and high-temperature steam corrosion of MAO/Cr bilayer coatings fabricated on Zr alloys. *J. Mater. Sci. Technol.* **2026**, *244*, 273–284. [CrossRef]
10. Cheng, B.; Zheng, G.; Hou, D.; Ding, C.; Chu, Q.; Feng, L.; An, G.; Zhang, X.; Sun, L.; Sun, H.; et al. Formation mechanisms of columnar/dense structure for $Gd_2Zr_2O_7$ceramics coatings with superior CMAS corrosion resistance. *Corros. Sci.* **2025**, *257*, 113348. [CrossRef]
11. Chen, M.; Jia, Z.; Wang, Z.; Xiang, C.; Pu, J.; Sun, G.; Han, E. Insights into the passivity and electrochemistry of CoCrFeMnNi high entropy alloy fabricated by underwater laser direct metal deposition. *Corros. Sci.* **2024**, *237*, 112289. [CrossRef]

materials

MDPI

Article

Tribological Properties of DLC Coatings in Model-Based and Real-Life Tests

Katarzyna Piotrowska [1], Monika Madej [1,*] and Krystian Milewski [2]

[1] Faculty of Mechatronics and Mechanical Engineering, Kielce University of Technology, Tysiąclecia Państwa Polskiego Ave., 25-614 Kielce, Poland; kpawelec@tu.kielce.pl
[2] TRZUSKAWICA S.A., 26-052 Sitkówka, Poland; krystian.milewski@trzuskawica.pl
* Correspondence: mmadej@tu.kielce.pl

Abstract

Machinery for internal transport in open-pit mines experiences excessive wear. Belt conveyors used for transporting aggregates are a type of equipment in which bearings are especially prone to failure. Considering the significant financial impact of equipment downtime, ensuring the high reliability of machinery in this sector is paramount. Consequently, the design of tribological interfaces should prioritize maximizing their reliability and minimizing the frequency of malfunctions. This article presents a comparative analysis of 100Cr6 steel and a-C:H type diamond-like carbon (DLC) coatings applied using chemical vapor deposition (PACVD) on bearing components in belt conveyors. Model-based tribological tests were conducted on these materials in both laboratory and real-life settings, evaluating friction and wear under dry friction and under Renolit UNI 3 grease-lubricated conditions, the latter being the operational lubricant for these bearings.

Keywords: bearings; belt conveyors; friction; DLC coatings; real-life tests

1. Introduction

The aggregate processing industry utilizes a range of mechanical systems, including crushers, screens, apron feeders, pulleys, and belt conveyors, all of which undergo significant operational wear [1]. Bearing elements in conveyors transporting materials over distances ranging from a dozen to several hundred meters are particularly susceptible to wear. Operational conveyor systems are exposed to abrasive stone dust generated from both the processing of raw materials and the transportation of crushed product, as well as to variations in temperature and humidity [2–4]. Even with the implementation of different sealing mechanisms, significant amounts of particulate matter from aggregates, prevalent in the operational surroundings, can infiltrate bearing interiors, thereby accelerating their abrasive degradation. The degradation of machinery components due to corrosion from contact with rainwater or process water used for aggregate rinsing presents a considerable risk. Furthermore, inadequate or incorrect lubrication contributes to bearing failure by increasing friction and causing overheating, which subsequently leads to pitting on the rolling surfaces and a reduction in operational lifespan. Vibrations and dynamic loads, particularly those caused by uneven conveyor loading or the transportation of course materials, also pose challenges. Extended exposure to these detrimental factors leads to the development of microcracks and material fatigue, resulting in machinery and equipment within the aggregate industry experiencing both tribological wear and other modes of deterioration [5–9]. Principally, each failure substantially contributes to

https://doi.org/10.3390/ma18184251

increased production costs. Figure 1 shows two types of damage in rolling bearings used in the aggregate industry.

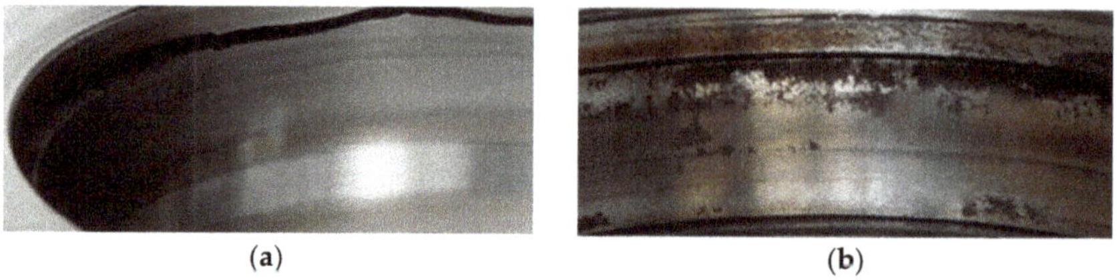

Figure 1. Photographs of the deterioration of rolling bearings used in the aggregate processing industry: (**a**) crack; (**b**) rust.

The adverse impacts of friction can be mitigated through the use of specialized lubricants and highly abrasion-resistant surface layers. Specifically chosen to meet the particular operational demands of a tribological system, they aim to reduce the friction coefficient, increase viscosity, enhance lubrication properties, and improve corrosion resistance [10].

The search for innovative solutions, driven by the rising costs of operating and maintaining industrial machinery and equipment, is a major factor in efforts to tailor diamond-like carbon coating properties for the harsh conditions typical in aggregate processing [11]. This adaptation has the potential to significantly improve the lifespan and reliability of machine components, which is vital for optimizing operational costs and enhancing the effectiveness of industrial procedures. Equipment breakdowns leading to production halts significantly impact a company's financial standing and reputation, particularly through customer order delays [12,13]. Consequently, the careful design of advanced tribological interfaces is crucial, especially for the aggregate sector, necessitating the selection of suitable anti-wear coatings and lubricants for specific friction point operating conditions. Furthermore, prior to field testing, it is essential to conduct appropriate laboratory experiments to provide an initial assessment of the effectiveness of the implemented solutions.

The materials and lubricants currently used in the working parts of machinery often fail to meet the growing demands of the aggregate processing industry, which operates under extreme conditions. Consequently, ongoing progress is essential through the development of more robust alloys, advanced protective coatings, and high-performance lubricants. This will lead to increased equipment lifespan and efficiency, while also reducing operational and maintenance expenses.

Anti-wear surface layers and coatings, such as diamond-like carbon (DLC) coatings applied in both lubricated and unlubricated friction nodes [8,14], have gained significant attention. High hardness and wear resistance, as well as a low coefficient of friction, make these coatings an exceptionally effective solution under conditions of intensive operation [15,16]. Their fundamental properties, mechanisms of action, and applications have been comprehensively reviewed in traditional works [17], while more recent reviews highlight the role of surface chemistry, hydrogen content, and tribochemical processes in achieving superlubricity in DLC films [18].

In [19], the authors compared the tribological properties of black iron oxide and WC/a-C:H coatings deposited on rolling bearings. The tests were performed on rolling bearings under laboratory conditions simulating environments with limited lubrication (which is typical in real mechanical applications). The obtained results indicated that WC/a-C:H coatings are more effective in protecting against adhesive wear and enhancing the operational durability of bearings under boundary lubrication conditions compared to the results for black oxide coatings. Additionally. WC/a-C:H coatings exhibited significantly lower wear and friction coefficients than those of black oxide. The authors of reference [20] analyzed the tribological behavior of ball bearings with a DLC coating deposited on both the inner and outer rings, focusing on the ball material and the effect of lubrication. The research outcomes indicated that pairing an a-C:H:Me coating with Si_3N_4 balls provided the most favorable performance, characterized by a consistent friction force throughout the duration of the experiment.

Furthermore, the use of DLC coatings in ball bearings eliminates the need for traditional mineral-oil-based lubricants, thereby contributing to energy and resource savings and enhancing the durability of bearings under dry friction conditions. The above studies provide valuable information on the potential applications of DLC coatings in industrial settings. The authors of [21] analyzed the influence of various gear oils on the wear of DLC coatings in gear applications. The research aimed to evaluate the effectiveness of the coatings under conditions similar to real-life conditions, taking into account various types of lubrication. The results of these studies indicate that the W-DLC/CrN coating effectively reduces abrasive wear and scuffing, particularly when used with PAO oil. However, it failed to improve pitting resistance, suggesting the need for further research on optimizing coatings for this type of damage.

DLC coatings were also investigated in the context of their application on rolling bearing components in the aerospace industry [22,23]. Pure hydrogenated DLC coatings were deposited on AISI 52100 steel. The authors aimed for high wear resistance coatings and a low coefficient of friction, both in vacuum and normal atmospheric environments. The tribological behavior was assessed in varying test environments (air, vacuum, and nitrogen) using a ball-on-disc tribometer, which employed AISI 52100 steel counter bodies. The measured mean coefficients of friction were in the ranges of 0.22–0.27 (in air), 0.02–0.03 (in nitrogen), and 0.007–0.013 (in vacuum). These results indicate a significant reduction in motion resistance achieved by the DLC coatings, especially in vacuum, positioning them as a potential material for tribological applications.

A comprehensive analysis of the effect of DLC coatings on the strength of 18CrNiMo7–6 steel gear teeth is presented in [24]. The study focused on strength and scuffing resistance tests. The results demonstrated that the DLC coating increased the surface hardness to approximately 2000 HV and the critical scuffing temperature from 224.6 °C to 348.6 °C, indicating improved resistance to extreme operating conditions. The findings of these studies suggest that DLC coatings significantly enhance the durability and reliability of machine components subjected to high loads and extreme temperatures. This is particularly important in the automotive, aerospace, and heavy machinery sectors, where mechanical components are intensely exposed to severe wear.

Recent research on the use of solid lubricants in combination with DLC has demonstrated that these solutions significantly improve the durability of the surface layer and enable the achievement of low friction coefficients in conditions of limited access to lubricating oils. Recent publications confirm this potential, demonstrating that appropriate functionalization of DLC with solid materials such as graphene, nanodiamonds, or hBN leads to the formation of stable transfer layers and enables the achievement of ultra-low friction—under both dry (<0.05) and humid (<0.1) conditions. However, some of the described methods require specific conditions, such as heat treatment in a vacuum. This clearly indicates that the integration of DLC with advanced solid lubricants is a promising development path in tribology [25–29].

Despite the significant interest in DLC coatings, the existing literature still lacks detailed studies regarding their application on aggregate processing equipment considering the specific operating conditions involving high abrasion, variable loads, and intensive wear. Most available research focuses on the use of DLC coatings in bearings, where their properties enable improved wear resistance, enhanced lubrication efficiency, and reduced energy losses. The aim of this work was to conduct a comparative analysis of how diamond-like carbon coatings influence the operational durability of bearings and to explore their potential application in belt conveyors.

2. Materials and Methods

The study investigated DLC coatings deposited on 100Cr6 steel using plasma-assisted chemical vapor deposition (PACVD) [30–32] with a glow discharge at a temperature below 250 °C. The layers were made using an RS 50 (Oerlikon Balzers, Pfäffikon, Switzerland) coating system. The selection of the above DLC coatings was based on the positive results of previous studies [1,6,7,11]. A schematic of the process is shown in Figure 2.

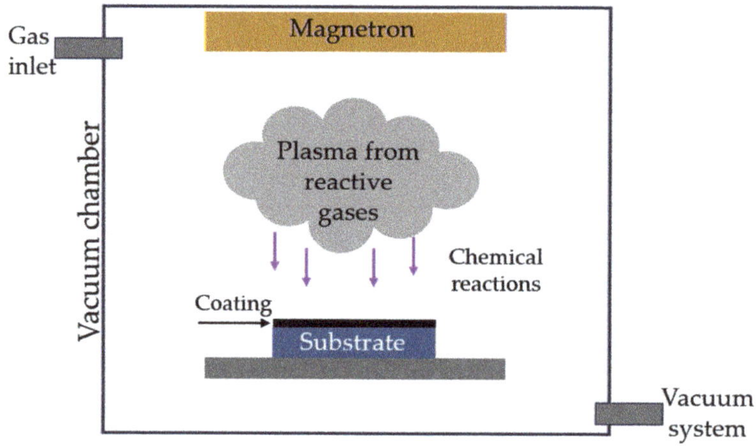

Figure 2. Schematic of the DLC coating deposition process—PACVD.

The thickness of the deposited coatings was examined via microscopic observations of the cross-sections using a JSM 7100F scanning electron microscope (JEOL, Tokyo, Japan) equipped with an EDS energy-dispersion spectrometer. An accelerating voltage of 15 kV and a magnification of ×20,000 were used. The study was extended to include point analyses of elemental composition in micro-areas. The test results are presented in Section 3.1.

Model-based tribotests were performed on both coated (a-C:H) and uncoated 100Cr6 steel discs (38 mm diameter, 6 mm height). The elemental composition of the 100Cr6 steel is detailed in Table 1. A TRB^3 tribometer (Anton Paar, Baden, Switzerland) was used in a rotary configuration under both dry and lubricated friction conditions. The test parameters are shown in Table 2, and the friction pair is shown in Figure 3.

Table 1. Chemical composition of 100Cr6 steel/wt. %.

Element	C	Mn	Si	P	S	Cr
Value	0.95–1.10	0.20–0.50	max–0.35	max–0.025	1.30–1.60	1.30–1.60

The samples were subjected to a normal load—F_N—of 10 N and 50 N, corresponding to the estimated initial contact pressures of around 1420 GPa and 2430 GPa for 100Cr6 and 1610 GPa and 2755 GPa for DLC.

Table 2. Tribological test parameters.

Parameter	Value and Unit
Type of movement	Rotary
Load	10 N and 50 N
Linear speed	0.1 m/s
Distance	1000 m
Friction radius	12 mm
Ambient temperature	23 ± 1 °C
Humidity	50 ± 1%
Lubricant	Without (DF), UNI 3
Lubricant temperature	23 ± 1 °C
Counter sample	100Cr6 (Ø 6 mm)

Figure 3. Friction pair.

For the lubricated tests, Renolit UNI 3 grease was used, a general-purpose grease composed of highly refined mineral oil and a lithium soap thickener (12-hydroxystearic acid), with selected properties listed in Table 3. During the tests, a layer of grease approximately 1 mm thick was applied to the sample. Following the tribological testing, the surface topography of the contact areas was analyzed. The results of the experiments are presented in Section 3.2.

Table 3. Selected properties of Renolit UNI 3 grease.

Property	Unit	Value
Color	-	Amber
Dropping point	°C	>190
Viscosity at 40 °C	mm^2/s	100
Operating temperature range	°C	−30 ÷ +130

Observations of the surface geometrical structure after the model-based tests were performed using a Leica DCM8 non-contact profilometer (Leica, Heerbrugg, Switzerland).

The analysis of the test results was carried out using LeicaMap 7.3 software. A 20× objective lens was used to characterize the operational surface layer. For each sample, a series of five measurements of a 1.2 mm × 1.2 mm area were captured in confocal mode. The surface topography analysis was conducted based on micrographs, 3D axonometric images, average transverse profiles of wear tracks, and the volumetric wear rate. The results of the experiment are presented in Section 3.3.

An additional experiment was conducted with DLC coating on a real bearing in service. The wear tracks on the outer raceway of the bearings, observed after tests under real-life conditions, were measured using a Form Talysurf PGI 1230 contact profilometer (Taylor Hobson, AMETEK, Inc., Berwyn, PA, USA). Due to the curved surface geometry of the bearings, which resulted in a large number of unmeasured points in the non-contact measurements, the contact method was employed to enable accurate mapping of the surface topography. Based on the obtained results, 3D axonometric images, average profiles of the wear tracks generated based on 50 profiles, and the most important amplitude parameters were prepared according to ISO 21920-2 [33]. The results are presented in Section 3.4.

3. Results and Discussion

3.1. Evaluation of DLC Coating Microstructure

Figures 4–6 present the morphology of the DLC coating. Specifically, Figure 4 illustrates the coating thickness, Figure 5 presents the chemical composition in selected micro-regions, and Figure 6 shows the linear elemental distribution.

Microstructure analysis revealed that the total coating thickness was approximately 2 μm, exhibiting a homogeneous structure across the entire examined surface. This suggests a stable deposition process and uniform coverage of the substrate, 100Cr6 steel. The homogeneity of the layer is significant from the perspective of its mechanical and tribological properties, as it minimizes the risk of localized stress concentrations that could lead to premature wear or delamination of the coating during operation.

The EDS analysis of the a-C:H coating in micro-areas revealed that the near-surface layer consisted of carbon and tungsten, while a chromium layer, acting as an interlayer, was identified at a depth of approximately 1.5 μm from the surface. The use of chromium was aimed at improving the adhesion of the coating to the metallic substrate.

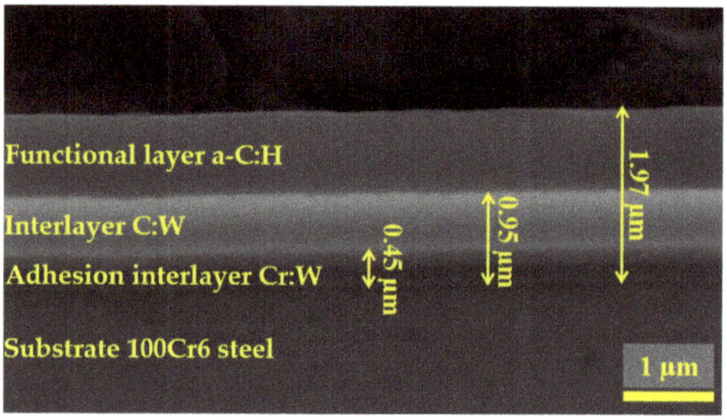

Figure 4. Morphology of the cross-section of the DLC coating.

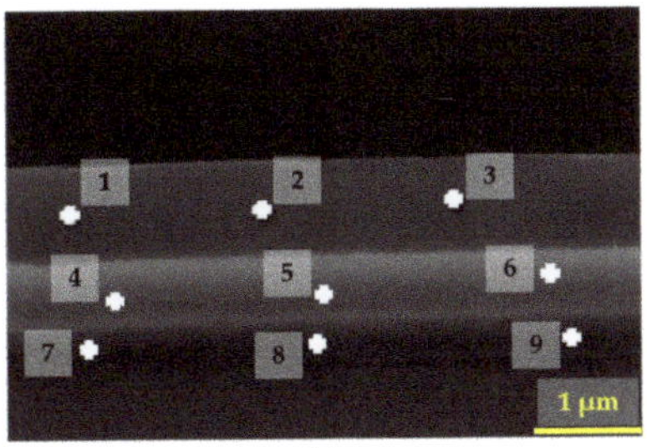

(a)

Spectrum	C	Cr	Fe	W
1	80.65	0.67	0.87	18.33
2	81.95	1.78	1.00	15.27
3	87.08	1.95	1.92	9.05
4	67.74	0.57	1.13	30.56
5	73.63	0.57	1.03	24.78
6	80.38	0.39	0.86	18.37
7	4.88	31.14	5.49	58.49
8	10.76	11.31	3.98	73.95
9	12.16	5.05	3.24	79.55

(b)

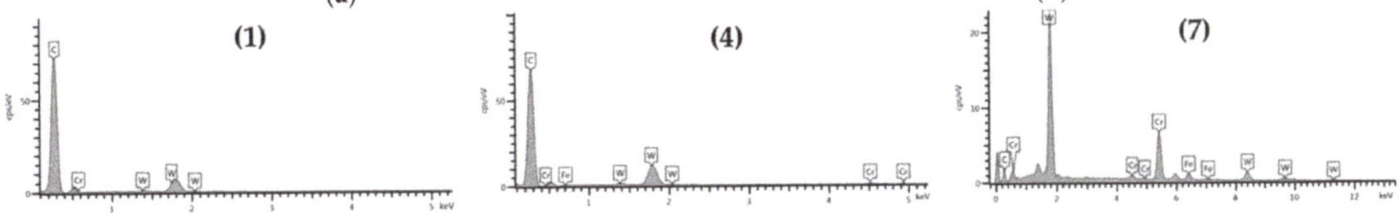

(c)

Figure 5. EDS analysis in micro-areas: (**a**) micrograph; (**b**) wt%; (**c**) characteristic radiation spectrum at points 1, 4, and 7.

3.2. Model-Based Tribological Tests

Figures 7 and 8 present the results of model-based tribotests conducted under dry (DF) and lubricated friction conditions using Renolit UNI 3 grease (UNI 3). The values of the friction coefficient and linear wear of the tested tribological pairs were recorded during the tests. Mean values and standard deviation of these parameters were determined based on three measurement series.

The deposition of a DLC coating enhanced the tribological performance of 100Cr6 steel under both 10 N and 50 N loads. Furthermore, lubrication improved operating conditions for both uncoated steel and the DLC-coated samples, with the most substantial benefits observed when the DLC coating and lubrication were combined.

Specifically, at a 10 N load, the DLC coating reduced the friction coefficient by nearly fivefold under technically dry conditions (0.70 to 0.15) and almost twofold with lubrication (0.21 to 0.11) when compared to the results for uncoated steel. Notably, tribological wear tests on the uncoated 100Cr6 steel–steel friction node at 50 N were not feasible due to a rapid increase in friction force (exceeding 25 N) that activated overload protection in the tribometer; this occurred during all three attempts.

However, tests on the DLC-coated samples at 50 N showed a reduction in motion resistance. Under dry conditions, the friction coefficient for the DLC coating increased slightly from 0.15 (at 10 N) to 0.19 (at 50 N). Still, with lubrication, it decreased significantly from 0.11 to 0.07, highlighting a synergistic effect with the lubricant, promoting low-shear sliding. Linear wear measurements confirmed that the DLC coating reduced the linear wear of 100Cr6 steel by approximately 68% under technically dry conditions and about 27% with lubrication. The results confirm that the DLC coating maintains low wear rates even under increased load.

The results show that at low load (10 N), lubrication significantly reduces the wear of 100Cr6 steel, while the DLC coating already provides low wear in dry conditions and exhibits an additional, although smaller, benefit with lubrication. At higher load (50 N), DLC wear remains nearly unchanged under both dry and lubricated conditions, demonstrating the coating's stability and resistance to load increase.

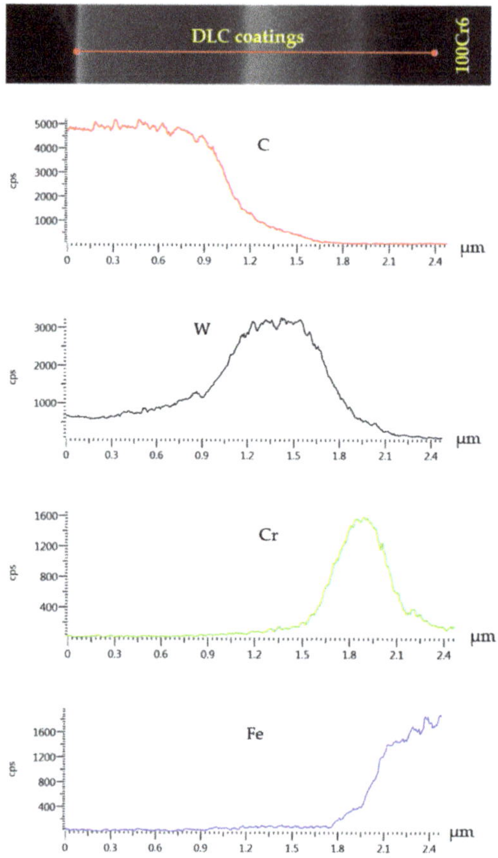

Figure 6. Linear distribution of elements on the cross-section.

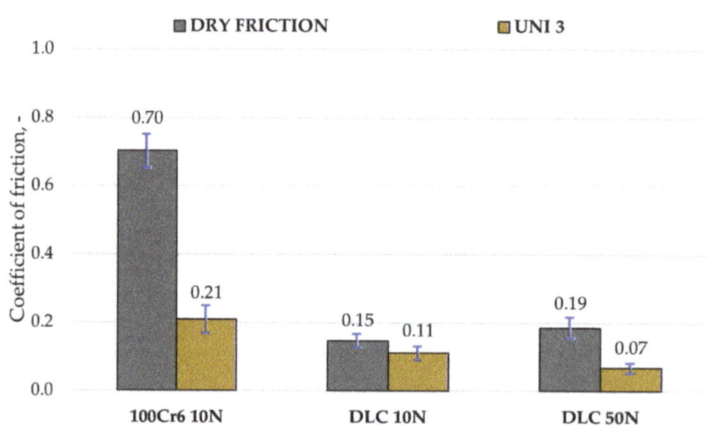

Figure 7. Mean values of friction coefficients obtained in model-based tests.

In conclusion, the combination of a DLC coating and lubrication yielded the best results for the tested tribopairs, achieving the lowest friction coefficients and linear wear. This suggests their promising potential as anti-wear solutions.

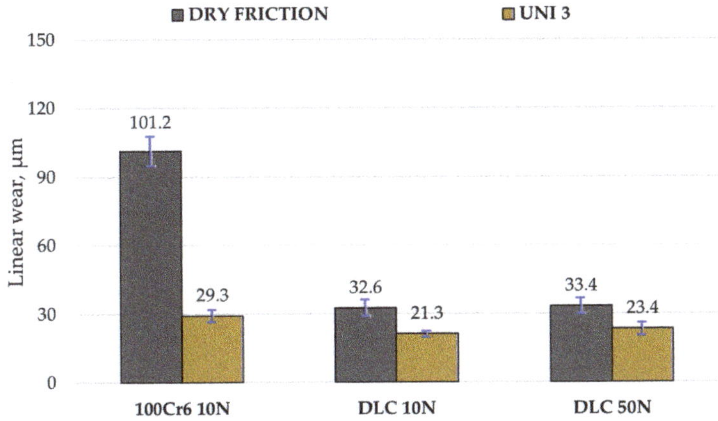

Figure 8. Mean linear wear of the tribological pairs obtained in model-based tests.

3.3. Tribological Wear Analysis Following the Model-Based Tests

After the model-based tests, the resulting wear tracks were subjected to microscopic observations. Photographs, 3D axonometric images, transverse profiles of the wear tracks, and volumetric wear rates calculated according to the following formula were analyzed [34]:

$$W_V = \frac{V}{F_N * s} \tag{1}$$

where:

V—volume of material removed, mm^3;

F_N—normal force, N;

s—distance, m.

The results are shown in Figures 9–14 and Table 4.

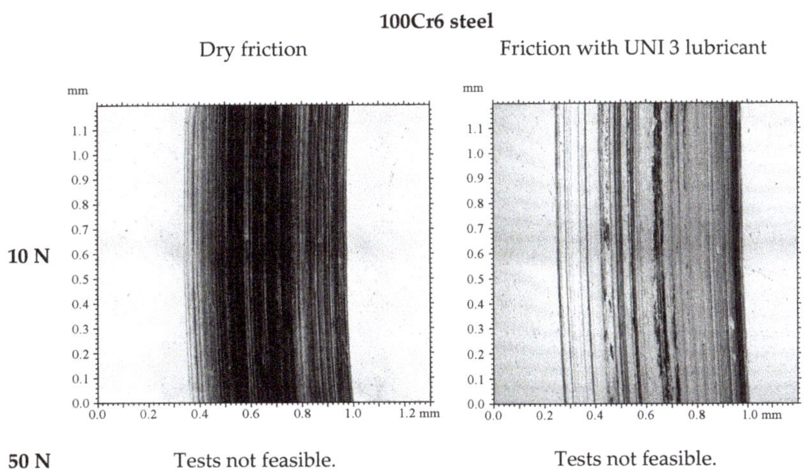

Figure 9. Photographs of wear tracks on 100Cr6 steel.

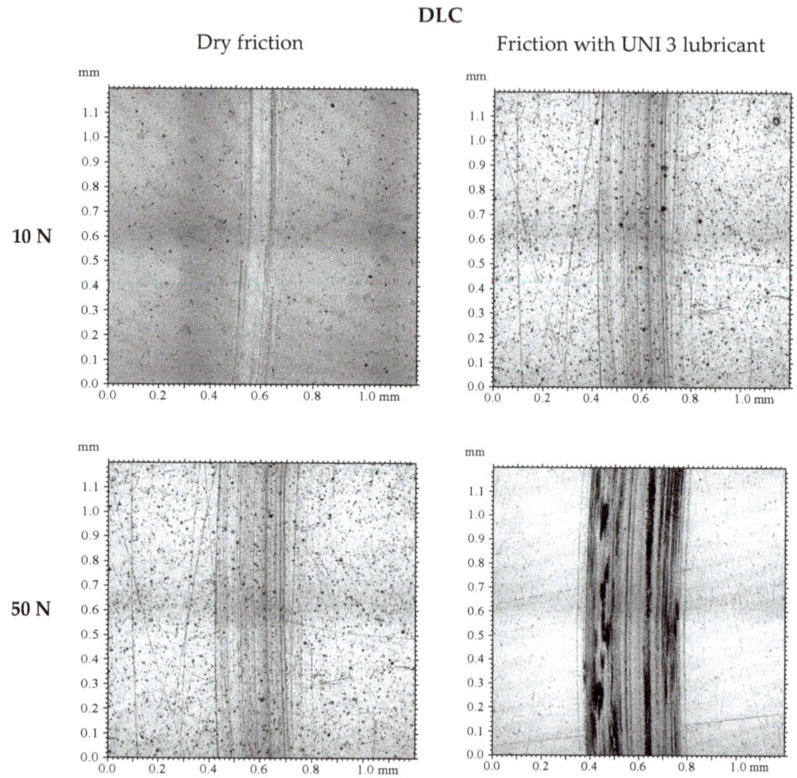

Figure 10. Photographs of wear tracks on the DLC coating.

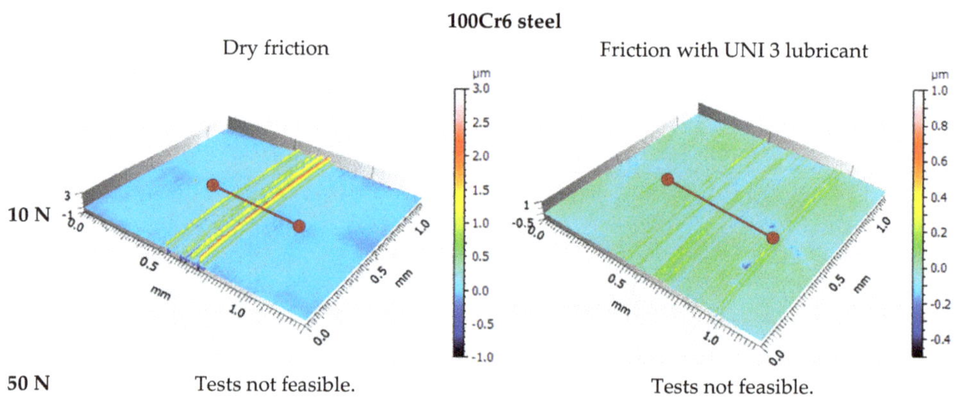

Figure 11. Axonometric views of wear tracks on 100Cr6 steel.

Analysis of sample wear following the model-based tests reveals that applying a DLC coating significantly decreased volumetric wear by 87% under technically dry conditions and by 77% with lubrication. Lubrication was most effective for the DLC coating at both 10 N and 50 N loads. Notably, the volumetric wear of the DLC coating remained relatively consistent when the load increased from 10 N to 50 N. While lubrication reduced wear for both 100Cr6 steel and DLC-coated samples, the effect was more pronounced for the DLC coating. These findings demonstrate that DLC coatings possess high resistance to increased loads and provide consistently low wear rates, positioning them as an excellent choice for components prone to excessive wear.

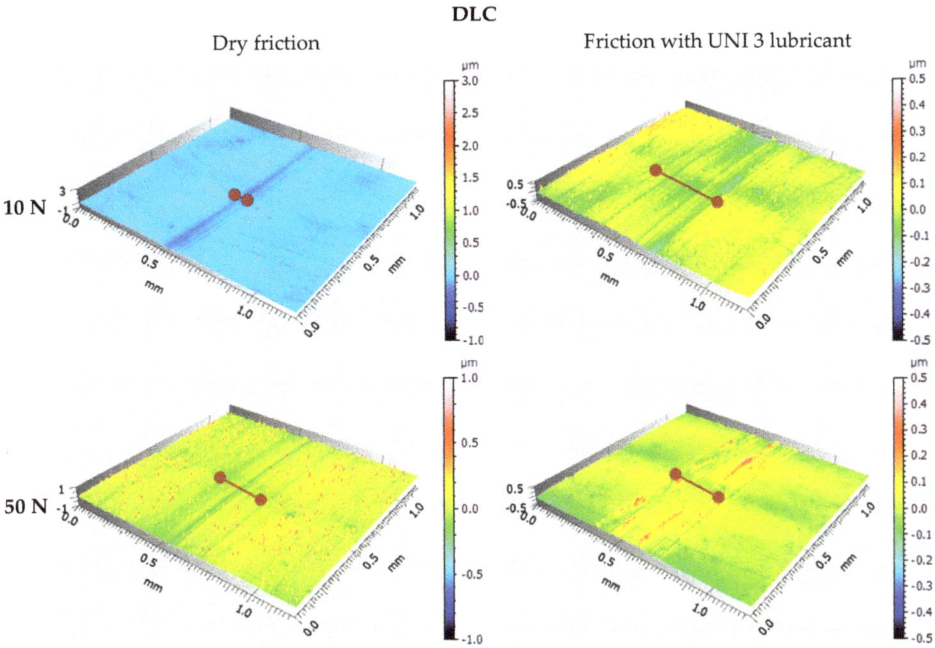

Figure 12. Axonometric views of wear tracks on the DLC coating.

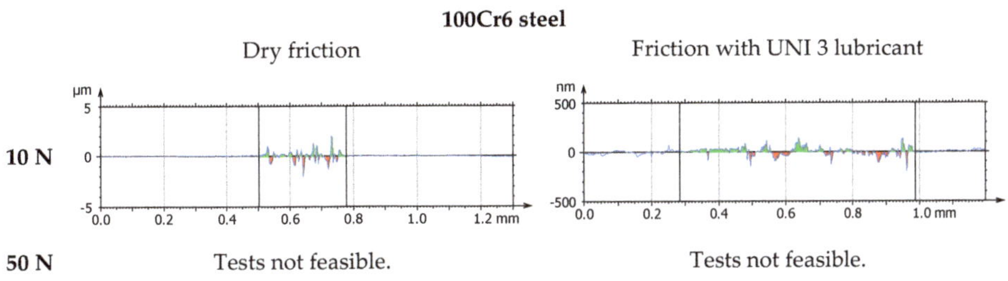

Figure 13. Transverse profiles of wear tracks on 100Cr6 steel.

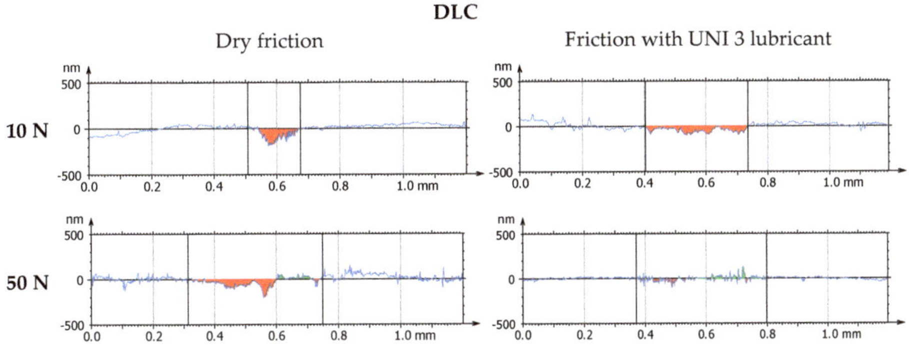

Figure 14. Transverse profiles of wear tracks on the DLC coating.

Table 4. Volumetric wear rates.

	Volumetric Wear Rate, $\left[\frac{mm^3}{N \times m}\right]$	
	10 N	**50 N**
100Cr6 DF	1.4×10^{-8}	-
100Cr6 UNI 3	3.9×10^{-9}	-
DLC DF	1.8×10^{-9}	2.0×10^{-9}
DLC UNI 3	8.9×10^{-10}	1.8×10^{-9}

3.4. Tribological Tests—Real-Life Conditions

For the real-life tests, a self-aligning ball bearing with a tapered bore, model 1217 K (ISO designation) from SKF (Svenska Kullagerfabriken, Poznań, Poland), was used. These bearings feature two rows of balls, a common spherical raceway in the outer ring, and two deep, uninterrupted raceway grooves in the inner ring. It has a tapered bore (suffix "K") and is supplied with normal radial internal clearance (C_0/C_N). Their design makes them insensitive to angular misalignment between the shaft and the housing, which can be caused by, for example, shaft deflection. The tapered bore allows for easy bearing mounting using an adapter or withdrawal sleeves. The dimensions are: 85 mm bore (d), 150 mm outside diameter (D), 28 mm width (B), shoulder diameter inner ring (d1), shoulder diameter outer ring (D_1) and min. 2 mm chamfer dimension ($r_{1,2}$). It is unsealed (no-seal) and uses a steel cage. The basic dynamic load rating (C) is approximately 48.8 kN, and the basic static load rating (C_0) is approximately 20.8 kN. The reference limiting speed is around 4700 rpm [35].

Components with and without DLC coatings were installed in the existing bearing housings of belt conveyors used for aggregate transport at the Aggregate Production and Dispatch Department of Trzuskawica S.A. Due to significant mechanical loads, intense material disintegration, and unfavorable environmental conditions—including corrosive effects—the operating conditions of these devices are classified as particularly harsh. Figure 15 shows a view of the bearing used for the real-life tests and its mounting location. It was assumed that, in the event of failure or after the end of their service life, the bearings were dismounted and submitted for technical evaluation and wear analysis.

(a) (b) (c)

Figure 15. Technical drawing of the bearing used for tests under real-life conditions [35] (**a**); photographs of bearing assembly locations (**b**,**c**).

3.5. Tribological Wear Analysis After Real-Life Tests

Figure 16 shows photographs of bearing segments after real-life tests: without coating (a), with a DLC coating (b). Figures 17 and 18 present the results of microscopic observation studies after real-life tests performed using contact profilometry. The 3D axonometric images (a) and wear track profiles in the cross-section (b) are compared. The wear analysis of bearings with and without a DLC coating was carried out based on three wear parameters: maximum wear track depth, cross-sectional area, and wear track volume (Table 5).

(**a**)

(**b**)

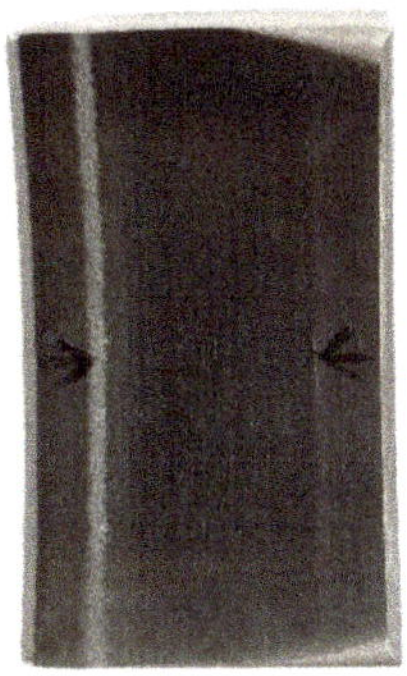

Figure 16. Photographs of bearing raceway segments after real-life condition tests: (**a**) without coating; (**b**) with a DLC coating.

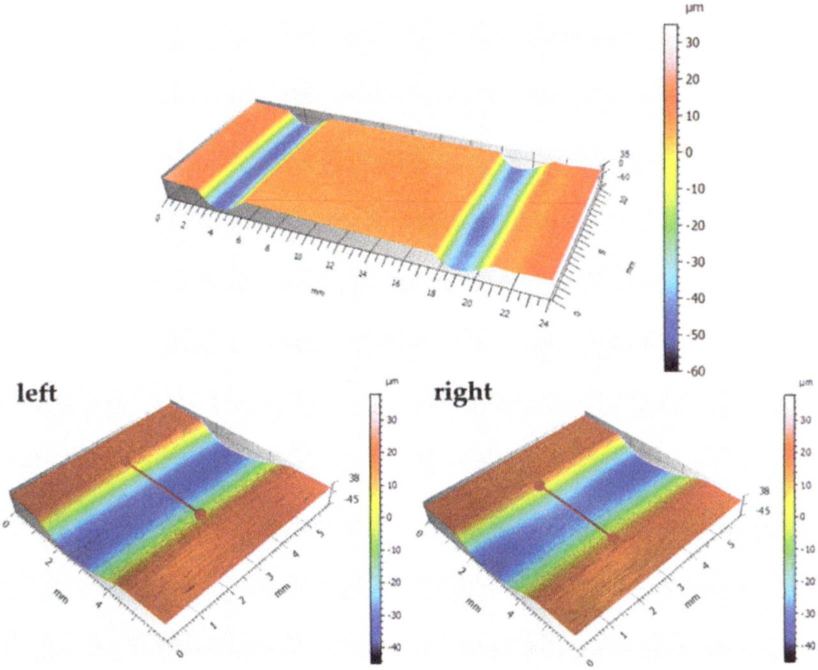

(**a**)

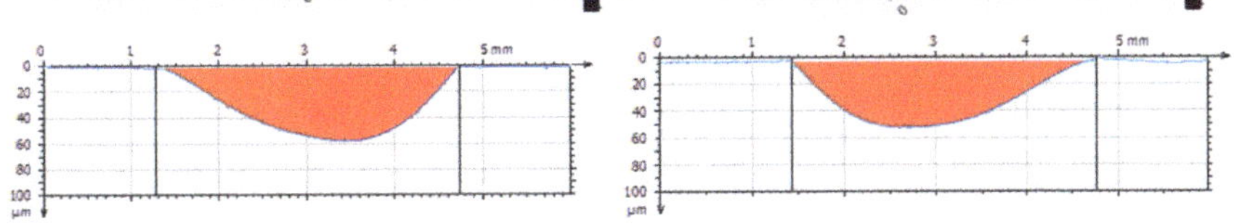

(**b**)

Figure 17. Wear analysis of a bearing without the coating: (**a**) 3D axonometric image; (**b**) wear profile.

Table 5. Bearing wear parameters.

		Max. Depth, mm	Surface Area, mm^2	Volume, mm^3
100Cr6	left	0.057	0.122	0.6498
	right	0.050	0.102	0.532
DLC	left	0.009	0.0057	0.0338
	right	0.001	0.0042	0.2180

After just six months of operation, the uncoated bearing failed and required replacement. However, the DLC-coated bearing continued to operate stably, showing no signs of critical damage. After 18 months, the decision was made to disassemble the bearing and conduct a wear assessment to verify the coating's actual effectiveness under industrial conditions.

The research results obtained within the scope of this work indicate that DLC coatings applied to the outer raceways of bearings contribute to the improvement of the tribological properties of the bearings, particularly in terms of reducing the friction coefficient and wear. This finding is confirmed by shallower wear track depths, smaller surface area, lower volumetric wear, and over three times longer operational life of the bearings with the DLC coating. Extending the service life of friction components in belt conveyors increases the operational availability of these machines in the aggregate industry.

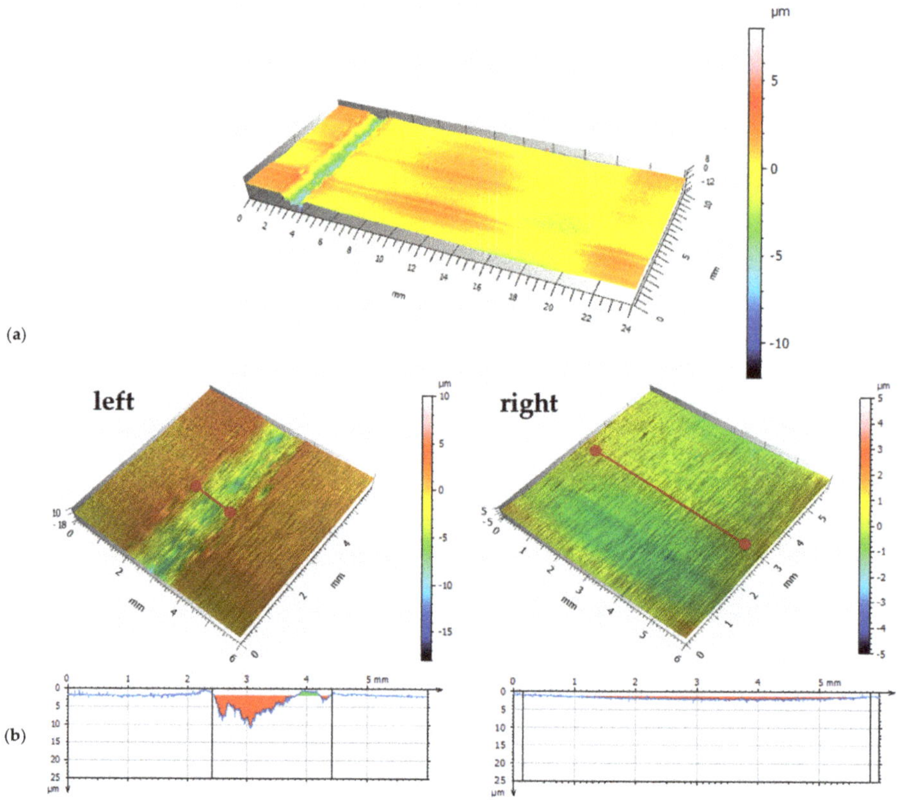

Figure 18. Wear analysis of a bearing with the DLC coating: (**a**) 3D axonometric image; (**b**) wear profile.

4. Conclusions

Model tests provide wear data, such as the coefficient of friction and volumetric wear, under controlled loads, lubrication times, and conditions. These results allow us to determine how the DLC coating reduces friction and wear compared to the results for uncoated steel. However, they do not directly predict bearing service life. The relationship between volumetric wear under laboratory conditions and bearing service life under real-world conditions is therefore qualitative, not quantitative. The laboratory tests with simplified contact geometry and controlled loads (10 and 50 N) demonstrate a low coefficient of friction and high wear resistance of the DLC coating under both dry friction and UNI 3 grease lubrication. Real-life tests take into account additional factors: dynamic loading, vibration, grease contamination, temperature gradients, and longer operating periods. These factors can lead to fatigue wear, pitting, and microcracks in the bearings. These conditions are not addressed in short-term tribological tests conducted in the laboratory.

Overall, laboratory tests demonstrate the potential of DLC coatings to reduce friction and wear, making them an exceptionally promising solution for bearing applications, but translating this into extended bearing service life requires additional, long-term testing.

Author Contributions: Conceptualization: K.P., M.M. and K.M.; methodology: K.P., M.M. and K.M.; writing—original draft preparation: K.P. and M.M.; writing—review and editing: K.P., M.M. and K.M., supervision, M.M. All authors have read and agreed to the published version of the manuscript.

Funding: This research received no external funding.

Institutional Review Board Statement: Not applicable.

Informed Consent Statement: Not applicable.

Data Availability Statement: The original contributions presented in this study are included in the article. Further inquiries can be directed to the corresponding author.

Conflicts of Interest: Krystian Milewski is employed by the TRZUSKAWICA S.A., 26-052 Sitkówka, Poland. The remaining authors declare that the research was conducted in the absence of any commercial or financial relationships that could be construed as potential conflicts of interest.

References

1. Milewski, K.; Piotrowska, K.; Madej, M. Assessment of the properties of diamond-like carbon coatings (DLC) used in the lime industry. *Metalurgija* **2023**, *63*, 81–84.
2. Xu, C.; Mudunuru, M.K.; Nakshatrala, K.B. Material degradation due to moisture and temperature. Part 1: Mathematical model, analysis, and analytical solution. *Contin. Mech. Thermodyn.* **2016**, *28*, 1847–1885. [CrossRef]
3. Industrial Magazine, The Most Demanding Bearing Applications. Available online: https://magazynprzemyslowy.pl/artykuly/najbardziej-wymagajace-zastosowania-dla-lozysk?utm_source=chatgpt.com (accessed on 22 June 2025).
4. Leshchynsky, V.; Wiśniewska-Weinert, H.; Magda, J.; Wiśniewski, T.; Rybak, T. A tribological device for tests in difficult exploitation conditions of bearing components with a modified surface layer. *Tribologia* **2010**, *4*, 283–292.
5. Kowalczyk, J.; Milewski, K.; Madej, M.; Ozimina, D. The properties of lubricated friction pairs with diamond-like carbon coatings. *Open Eng.* **2020**, *10*, 688–698. [CrossRef]
6. Kuznetsova, T.; Lapitskaya, V.; Khabarava, A.; Trukhan, R.; Chizhik, S.; Torskaya, E.; Fyodorov, S.; Aizikovich, S.; Sadyrin, E.; Warcholinski, B. Features of wear of DLC-Si coating under microcontact conditions during the formation of secondary structures. *Compos. Struct.* **2023**, *316*, 117039. [CrossRef]
7. Milewski, K.; Madej, M.; Ozimina, D. Tribological properties of diamond-like carbon coatings at friction joints lubricated with ionic liquid. *Tribologia* **2019**, *5*, 59–69. [CrossRef]
8. Sulaiman, M.H.; Farahana, R.N.; Mustaffa, M.N.; Bienk, K. Tribological properties of DLC coating under lubricated and dry friction condition. *Mater. Sci. Eng.* **2019**, *670*, 012052. [CrossRef]

9. Soprano, P.B.; Salvaro, D.B.; Giacomelli, R.O.; Binder, C.; Klein, A.N.; Basoli de Mello, J.D. Effect of soft substrate topography on tribological behavior of multifunctional DLC coatings. *J. Braz. Soc. Mech. Sci. Eng.* **2018**, *40*, 371. [CrossRef]
10. Madej, M. *Properties of Tribological Systems with Diamond-Like Coatings, Wyd*; Politechniki Świętokrzyskiej, Kielce: Kielce, Poland, 2013.
11. Milewski, K.; Kudliński, J.; Pająk, M.; Madej, M.; Ozimina, D. Zastosowanie powłok diamentopodobnych na element trące środków transport wewnętrznego w przemyśle kruszywowym. *Autobusy* **2016**, *10*, 150–156.
12. West, J.; Siddhpura, M.; Evangelista, A.; Haddad, A. Improving Equipment Maintenance—Switching from Corrective to Preventative Maintenance Strategies. *Buildings* **2014**, *14*, 3581. [CrossRef]
13. Available online: https://www.makula.io/blog/downtime-in-manufacturing?utm_source=chatgpt.com (accessed on 15 June 2025).
14. Mo, J.L.; Zhu, M.H. Tribological investigation of WC/C coating under dry slidingconditions. *Wear* **2011**, *271*, 1998–2005. [CrossRef]
15. Tuszyński, W.; Michalczewski, R.; Osuch-Słomka, E.; Snarski-Adamski, A.; Kalbarczyk, M.; Wieczorek, A.; Nędza, J. Abrasive Wear, Scuffing and Rolling Contact Fatigue of DLC-Coated 18CrNiMo7-6 Steel Lubricated by a Pure and Contaminated Gear Oil. *Materials* **2021**, *14*, 7086. [CrossRef]
16. Pancielejko, M.; Czyżniewski, A.; Gilewicz, A.; Zavaleyev, V. Properties of DLC coatings with chromium-based sublayers used for woodworking high-speed steel tools. *Inżynieria Mater.* **2014**, *35*, 535–539.
17. Donnet, C.; Erdemir, A. *Tribology of Diamond-Like Carbon Films. Fundamentals and Applications*; Springer Science & Business Media: Berlin, Germany, 2008; Available online: https://link.springer.com/book/10.1007/978-0-387-49891-1 (accessed on 22 June 2025).
18. Erdemir, A.; Eryilmaz, O. Friction: Achieving superlubricity in DLC films by controlling bulk, surface, and tribochemistry. *Friction* **2014**, *2*, 140–155. Available online: https://link.springer.com/article/10.1007/s40544-014-0055-1 (accessed on 22 June 2025). [CrossRef]
19. Evans, R.D.; Hager, C.H., Jr.; Kang, Y.S.; Doll, G.L. Comparison of Black Oxide and Tungsten Carbide–Reinforced Diamond-Like Carbon (WC/a-C: H) Surface Treatments for Rolling Element Bearings. *Tribol. Trans.* **2015**, *58*, 444–453. [CrossRef]
20. Kröner, J.; Kursawe, S.; Musayev, Y.; Tremmel, S. Analysing the Tribological Behaviour of DLC-Coated Dry-Running Deep Groove Ball Bearings with Regard to the Ball Material. *Appl. Mech. Mater.* **2015**, *856*, 444–453. [CrossRef]
21. Michalczewski, R.; Kalbarczyk, M.; Mańkowska-Snopczyńska, A.; Osuch-Słomka, E.; Piekoszewski, W.; Snarski-Adamski, A.; Szczerek, M.; Tuszyński, W.; Wulczyński, J.; Wieczorek, A. The Effect of a Gear Oil on Abrasion, Scuffing, and Pitting of the DLC-Coated 18CrNiMo7-6 Steel. *Coatings* **2019**, *9*, 2. [CrossRef]
22. Xiufang, L.; Jibin, P.; Liping, W.; Qunji, X. Novel DLC/ionic liquid/graphene nanocomposite coatings towards high-vacuum related space applications. *J. Mater. Chem. A* **2013**, *1*, 3797–3809. [CrossRef]
23. Vanhulsela, A.; Velascobc, F.; Jacobsa, R.; Eerselsa, L.; Havermansa, D.; Robertsb, E.W.; Sherringtonc, I.; Andersonb, M.J.; Gaillardd, L. DLC solid lubricant coatings on ball bearings for space applications. *Tribol. Int.* **2007**, *40*, 1186–1194. [CrossRef]
24. Wu, J.; Wei, P.; Liu, G.; Chen, D.; Zhang, X.; Chen, T.; Liu, H. A comprehensive evaluation of DLC coating on gear bending fatigue, contact fatigue, and scuffing performance. *Wear* **2024**, *536*, 205177. [CrossRef]
25. Berman, D.; Deshmukh, S.A.; Sankaranarayanan, S.K.R.S.; Erdemir, A.; Sumant, A.V. Macroscale superlubricity enabled by graphene nanoscroll formation. *Science* **2015**, *348*, 1118–1122. [CrossRef]
26. Berman, D.; Narayanan, B.; Cherukara, M.J.; Sankaranarayanan, S.K.R.S.; Erdemir, A.; Zinovev, A.; Sumant, A.V. Operando tribochemical formation of onion-like-carbon leads to macroscale superlubricity. *Nat. Commun.* **2018**, *9*, 1164. [CrossRef] [PubMed]
27. Huang, P.; Chen, X.; Qi, W.; Tian, J.; Xu, J.; Wang, K.; Deng, W.; Zhang, C.; Luo, J. Vacuum Thermal Treatment for Achieving Macroscale Superlubricity by Nanodiamond and Hexagonal Boron Nitride on H-DLC Film Surfaces in Dry Nitrogen. *Adv. Funct. Mater.* **2024**, *34*, 49. [CrossRef]
28. Mescola, A.; Lodi, A.; Zanni, F.; Rota, A.; Gerbi, A.; Bernini, C.; Schott, M.; Repetto, L.; Camisasca, A.; Giordani, S.; et al. Synergistic effect of graphene and nanodiamonds to achieve ultra-low friction on rough DLC coatings. *Diam. Relat. Mater.* **2024**, *145*, 111149. [CrossRef]
29. Mescola, A.; Bertoni, G.; Gazzadi, G.C.; Bartkowski, M.; Camisasca, A.; Giordani, S.; Buzio, R.; Paolicelli, G. Humidity-Resistant Ultralow Friction in Diamond-Like Carbon Coatings Enabled by Graphitic Nanodiamonds. *Small Struct.* **2025**, 2500236. [CrossRef]
30. Sharifahmadian, O.; Mahboubi, F.; Oskouie, A. Structural evolution and tribological behavior of nitrogen-doped DLC coatings deposited by pulsed DC PACVD method. *Diam. Relat. Mater.* **2019**, *91*, 74–83. [CrossRef]
31. Al-Asadi, M.M.; Al-Tameemi, H.A. The effect of diamond like carbon coating on the wear resistance at dry sliding conditions. *Mater. Res. Express* **2022**, *9*, 116504. [CrossRef]
32. Marin, E.; Lanzutti, A.; Nakamura, M.; Zanocco, M.; Zhu, W.; Pezzotti, G.; Andreatta, F. Corrosion and scratch resistance of DLC coatings applied on chromium molybdenum steel. *Surf. Coat. Technol.* **2019**, *378*, 124944. [CrossRef]
33. *ISO 21920-2:2021*; Geometrical Product Specifications (GPS)—Surface Texture: Profile—Part 2: Terms, Definitions and Surface Texture Parameters. International Organization for Standardization: Geneva, Switzerland, 2021.

34. Kot, M.; Rakowski, W.; Major, Ł.; Lackner, J. Examination of the fracture toughness and wear resistance of multilayer coatings. *Tribologia* **2014**, *2*, 89–99.
35. Available online: https://www.skf.com/pl/products/rolling-bearings/ball-bearings/self-aligning-ball-bearings/productid-1217%20K (accessed on 22 June 2025).

materials

MDPI

Article

Preliminary Study on Electrochemical Deposition of Graphene on Steel Substrate via In Situ Oxidation Using Cyclic Voltammetry

Mattia Pelucchi [1,*], Brigida Alfano [2,*], Giuseppe Cesare Lama [3], Raphael Palucci Rosa [1] and Marina Cabrini [1,4,5,*]

[1] Department of Engineering and Applied Sciences, University of Bergamo, 24044 Dalmine, Italy; raphael.rosa@unibg.it
[2] TERIN-FSD-DIN, ENEA—Italian National Agency for New Technologies, Energy and Sustainable Economic Development, R.C. ENEA, 80055 Portici, Italy
[3] Institute of Polymers, Composites and Biomaterials, National Research Council (IPCB-CNR), 80055 Portici, Italy; giuseppecesare.lama@cnr.it
[4] National Interuniversity Consortium of Materials Science and Technology (INSTM), 50121 Firenze, Italy
[5] Research Unit of Bergamo, CSGI—Centre of Colloid and Surface Science, 22044 Dalmine, Italy
[*] Correspondence: mattia.pelucchi@unibg.it (M.P.); brigida.alfano@enea.it (B.A.); marina.cabrini@unibg.it (M.C.)

Abstract: This study explores an innovative method for depositing graphene directly onto metal surfaces, using cyclic voltammetry with a suspension of graphene in water. Most electrochemical deposition techniques up to now have concentrated on graphene oxide (GO) rather than pure graphene, largely because GO disperses more readily in water. This characteristic makes GO simpler to manipulate and apply in deposition processes, giving it an advantage in terms of usability and practicality. We demonstrated that graphene can indeed be deposited onto metal surfaces using this innovative electrochemical approach. We conducted a thorough characterization of the resulting graphene deposits, employing advanced techniques, including interferometric microscopy, scanning electron microscopy (SEM), energy-dispersive X-ray spectroscopy (EDS), and Raman spectroscopy. These analyses provided us with insights into the structural, chemical, and morphological characteristics of the graphene deposits. This comparison allowed us to assess the strengths and potential improvements needed for this direct deposition method, as it offers a more sustainable and streamlined alternative to conventional GO-based processes. One important finding is that, while the quality of these direct graphene deposits has not yet reached the level of GO-based coatings, this new approach has some compelling advantages. Specifically, it is a simpler, more environmentally friendly process that could streamline production and reduce the environmental impact compared to traditional methods using GO.

Keywords: graphene coating; cyclic voltammetry deposition; in situ oxidation of graphene

1. Introduction

The use of hydrogen as an energy carrier with a low environmental impact is considered a cornerstone of the energy transition, so much so that the European Union has included hydrogen production among the necessary steps for the decarbonization of the continent by 2050 [1]. At present, electrolytic hydrogen production can be convenient for exploiting peaks of overproduction of electricity from renewable sources such as wind and solar power as an alternative to storage in chemical batteries [2,3]. The hydrogen thus produced is blended with natural gas and fed into the gas distribution network.

In the dossier 'Blending hydrogen from electrolysis into the European gas grid' published by the European Commission's Joint Research Centre (JRC), aimed at analyzing the impact of a future hydrogen injection into the European gas grid, it is reported that the current natural gas distribution grids can immediately accommodate a 5% share of H_2, while by 2030 this share could be increased to 20% [4]. The problems associated with the transport of hydrogen in natural gas distribution networks currently in use are due to the possibility of dissociation of the hydrogen molecule on pipe walls that are made of steel, followed by diffusion into the metallic material with the possibility of embrittlement occurring [5]. Various methods proposed to prevent hydrogen from entering metallic materials include the application of polymeric [6], metallic [7], and ceramic coatings [8–10].

The application of multilayer polymeric coatings, such as fusion-bonded epoxy (FBE), polyethylene (PE), and multilayer polypropylene (PP) systems, is a key part of state-of-the-art industrial practice for mitigating hydrogen-induced degradation in natural gas pipelines [11]. These coatings act as primary barriers, limiting contact between moisture and aggressive species (including molecular hydrogen and hydrogen sulfide) and the wall pipe [11,12]. Moreover, the use of metallic underlayers (e.g., zinc-rich primers) is sometimes a feature of improved adhesion and enhanced cathodic protection. However, traditional coatings often have limitations regarding long-term hydrogen impermeability, particularly under the elevated pressure and temperature conditions that may be present in future hydrogen-enriched gas networks. Interest in alternative barrier technologies, including nanostructured coatings and two-dimensional materials like graphene, has therefore increased, as these are aimed at providing enhanced protection against hydrogen permeation and associated embrittlement phenomena [13]. The use of graphene as a coating potentially capable of reducing hydrogen ingress into metals is of interest both for pipelines and also for small components like regulation valves, sensors, and measurement dispositive.

Graphene, a two-dimensional sheet of sp^2-hybridized carbon atoms, has recently garnered significant attention due to its unique combination of electrical, mechanical, and optical properties. These exceptional characteristics make graphene a promising material for a broad range of applications, including electronics, sensors, composites, and corrosion protection [14,15]. In addition, graphene is also impermeable; its dense lattice structure forms an effective barrier that blocks even helium atoms. This remarkable property makes graphene an ideal candidate for applications requiring high levels of impermeability [16–18]. Nam et al. studied the applicability of graphene deposited by chemical vapor deposition (CVD) on copper to prevent hydrogen embrittlement, finding a marked improvement up to 25% deformation of the copper [19]. The protective effect was attributed to the formation of a C-H bond that blocks hydrogen from entering the metal [19]. Shi et al. realized multilayer graphene deposits using ionic implantation on a nickel deposit, followed by high-temperature solubilization, demonstrating that the coating is able to decrease the corrosion rate of an X70 steel and slow down the diffusion of hydrogen [20].

However, these techniques require specific machinery capable of vacuuming the chamber where the process takes place and controlling the rate of evaporation and deposition of carbon on the substrate to give graphene. These techniques appear too slow and too expensive to be applied on a large and industrial scale for the protection of carbon steel in contact with hydrogen gas. Although the application of a graphene coating on pipes is not currently feasible, it could be performed on small components of complex shapes that would be difficult to coat with high vacuum techniques [21]. Currently, high-coverage deposits are primarily limited to small surface areas and typically require expensive chemical vapor deposition (CVD) processes [22,23]. Despite the recent attainment of CVD coatings measuring tens of millimeters squared [24], for larger surfaces the most reliable method involves the electroplating of chemically oxidized graphene (GO) because it is simpler, requires less equipment, and is more cost-effective [25–27]. Kim et al. [28] successfully deposited graphene on stainless steel using a GO solution, applying a constant potential of 10 V for 8 min. Their research aimed to evaluate the barrier effect of these deposits on hydrogen embrittlement, an important consideration in materials science. Similarly, Marcin Behunová et al. [29] achieved consistent and homogeneous graphene deposits on stainless steel specimens using the same voltage and a minimum duration of 8 min. Currently, graphene oxide (GO) is produced through the chemical oxidation of graphite using the Hummers method [24], which employs aggressive reagents that have a considerable environmental impact.

The aim of this work is to bypass the conventional step of converting graphite into graphene oxide (GO) through chemical oxidation. Instead, this method uses an electrochemical technique that enables in situ oxidation and subsequent rapid reduction directly on the metal surface. This approach simplifies the deposition process, making graphene coatings more accessible, environmentally sustainable, and well-suited for practical applications. Eliminating the need for chemical oxidation reduces both the complexity and environmental impact of graphene production, paving the way for broader use in industrial and anticorrosive coatings.

The relevance of this work lies in the fact that, in the literature, numerous articles address graphene deposition starting from chemically obtained graphene oxide and applying a constant potential [25–29], whereas only a few studies focus on direct electrochemical oxidation [30] and deposition of graphene.

2. Materials and Methods

2.1. Solutions

Two types of graphene-based solutions were utilized: the first was a solution of graphene oxide (GO) at a concentration of 1 mg/mL in distilled water, synthesized via the traditional Hummers method [31]. This solution will later be referred to as the "GO solution". The second solution consisted of graphene flakes dispersed in a 50% mixture of distilled water and ethanol, at a concentration of 1 mg/mL, which will later be referred to as "GE solution". This suspension required sonication and the addition of 2 g/L of sodium lauryl sulfate to reduce surface tension, thereby enhancing the stability and homogeneity of the suspension. To ensure dispersion and prevent the aggregation of graphene flakes, both solutions were subjected to a two-hour sonication period prior to every test.

Graphene flakes that were used for the GE solution were synthesized using the liquid phase exfoliation (LPE) technique. This technique involves dispersing graphite flakes in a hydro-alcoholic solution and then sonicating the solution, which induces mechanical shear forces that separate the graphene layers from the graphite structure, resulting in high-quality, few-layer graphene flakes [32]. Afterwards, graphene flakes were separated from the solvents using freeze-drying. Graphene flakes thus obtained will be later referred to as "GE powder" (graphene-exfoliated powder).

2.2. Specimens

Wire-cut specimens of API 5L X65 pipeline steel were obtained from a gas pipeline. This steel is commonly used in the piping industry, and its chemical composition is shown in Table 1. Steels with similar compositions should produce the same results in terms of graphene deposits as API 5L X65 because the latter is typically composed of low-alloy carbon steel. The specimens were cylindrical with a diameter of 15 mm and a thickness of 5 mm, which was required to enable the use of a specimen holder compatible with the standard ASTM G5 cell [33]. All specimens underwent a surface preparation protocol involving progressive polishing with abrasive papers up to 400 grit to achieve a uniform finish. To ensure the removal of organic contaminants, each specimen was subjected to triple sequential washing with soap, followed by immersion in acetone within an ultrasonic bath. Prior to deposition, steel surfaces were activated via brief pickling in inhibited hydrochloric acid to activate surface reactivity and eliminate any oxide layers.

Table 1. Mass fraction composition of API 5L X65 steel.

Weight %	C max	Mn max	P max	S max
API 5L X65	1.45	0.03	0.03	0.03

2.3. Potentiostatic Deposition

This method involves the deposition of graphene and graphene oxide onto metal substrates through a controlled electrochemical process, applying a constant potential over a set period as described in previous studies [25–29]. The deposition process was carried out at room temperature within a glass electrochemical cell containing the specimen as the working electrode, an activated titanium counter electrode, and a AMEL 390/TCG saturated calomel reference electrode with a potential of +0.240 V vs. NHE at 25 °C (Figure 1). The reference electrode was positioned as close to the working electrode as possible to minimize ohmic drops in the solution. Throughout the experiments, the separation between the steel specimen and the counter electrode was maintained at 20 mm. A potential of 10 V was applied for a deposition duration of 8 min, with the specimen subjected to anodic polarization. The potentiostat employed for the experiments was an AMEL 7050 (AMEL s.r.l, Milan, Italy). These parameters align with those frequently cited in literature [28,29].

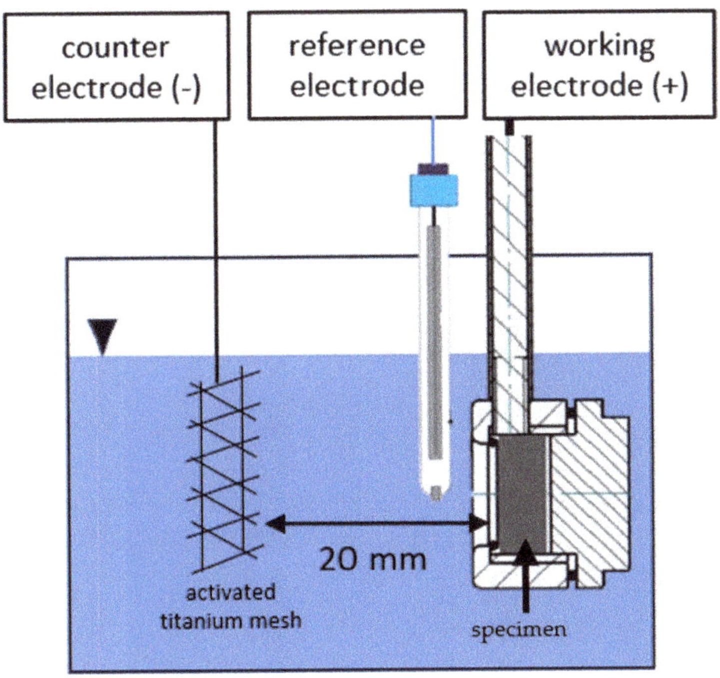

Figure 1. Potentiostatic cell setup.

2.4. Cyclic Voltammetry Deposition

The graphene solution used in these tests was identical to that used in potentiostatic experiments (GE solution), following the same preparation methodology. To minimize corrosion of the metallic substrate, the solution pH, initially at approximately 5.5, was adjusted to pH 13 by adding sodium hydroxide, as further detailed in the discussion section. The specimens utilized for the cyclic voltammetry tests were identical to those prepared for the potentiostatic deposition, having undergone identical surface preparation procedures. This ensured comparability of the results across both experimental setups. The deposition was conducted at room temperature using a classic three-electrode voltammetry setup [34,35]. In this configuration, the working electrode was the specimen, the counter electrode was a graphite rod, and the reference electrode was an AMEL 390/TCG saturated calomel electrode with a potential of +0.240 V vs. NHE at 25 °C. The reference electrode was positioned as close to the working electrode as possible to minimize ohmic drops in the solution. Throughout the experiments, the separation between the steel specimen and the counter electrode was maintained at 20 mm (Figure 2). During these tests, the parameters investigated included the applied potential range and the number of cycles. The potentiostat employed for the experiments was an AMEL 7050 (AMEL s.r.l, Italy), programmed to perform cyclic voltammetry at a scanning rate of 50 mV/s, varying the potential according to a triangular waveform across a specified range depending on the test conditions. Following the deposition process, the specimen was rinsed with distilled water and subsequently dried in a laboratory dryer before undergoing analysis.

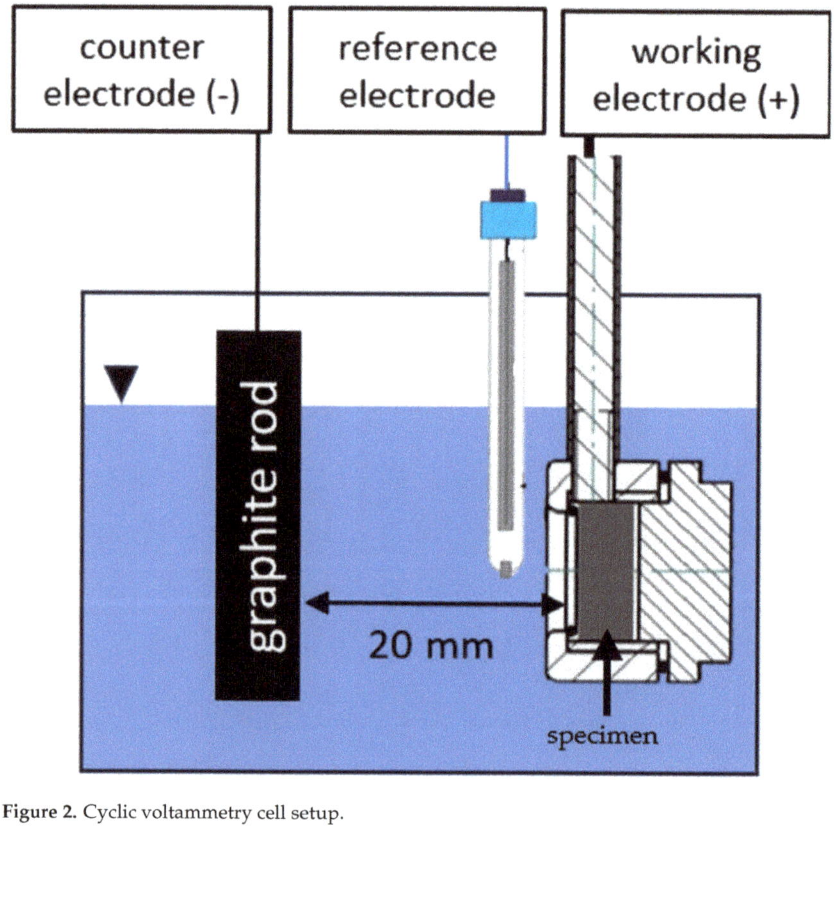

Figure 2. Cyclic voltammetry cell setup.

2.5. Coating Characterization Methods

The deposits were initially characterized by being observed under an optical microscope. The surface of deposits, which appeared to be best in terms of uniformity and degree of coating, was reconstructed using an interferometric microscope to assess the surface roughness and to evaluate the shape and depth of any craters. These deposits were also observed using a SEM equipped with an EDX probe. Samples were also analyzed using a Raman spectroscopy system.

2.6. Instruments and Software Used

The digital microscope utilized in this study was a KEYHENCE VHX-7000 (KEYHENCE ITALIA spa, Milan, Italy). The interferometric microscope utilized in this study was a SENSOFAR SNEOX (Sensofar, Terrassa, Spain) interferometric microscope. SEM analyses were done using a ZEISS SIGMA 300 (Carl Zeiss S.p.A., Milan, Italy) field emission scanning electron microscope (FESEM) equipped with an OXFORD XACT energy-dispersive x-ray spectroscopy (EDX) probe. The parameters utilized for the acquisition of SEM photographs are specified in the caption of each image. The EDX spectra were produced using a beam acceleration voltage of 8 keV, as the heaviest element in the specimens analyzed was iron.

Samples were also analyzed using a HORIBA XploRA PLUS Raman spectroscopy system (HORIBA ITALIA SRL, Rome, Italy) with a 532 nm laser; the acquired spectra are the result of 10 accumulations, performed in backscattering configuration. In order to avoid overheating the deposit, the laser was used at 1% of its maximum power (P_{max} = 20 mW).

The data were processed and subsequently represented in the graphs of this article using an electronic spreadsheet, i.e., Excel.

3. Results and Discussion

3.1. GE Powder

Before the dispersion, graphene flakes (GE powder) were analyzed using scanning electron microscopy (SEM) and energy-dispersive X-ray (EDX). GE powder is made of particles of different shapes and sizes and is not homogeneous (Figure 3); EDX analysis shows that it is composed essentially of carbon (95% atomic) and some oxygen (3% atomic); traces of aluminum and silicon are also present (Table 2). The carbon to oxygen ratio is very high (~31.7), and the presence of impurities such as Al and Si is probably due to the production process.

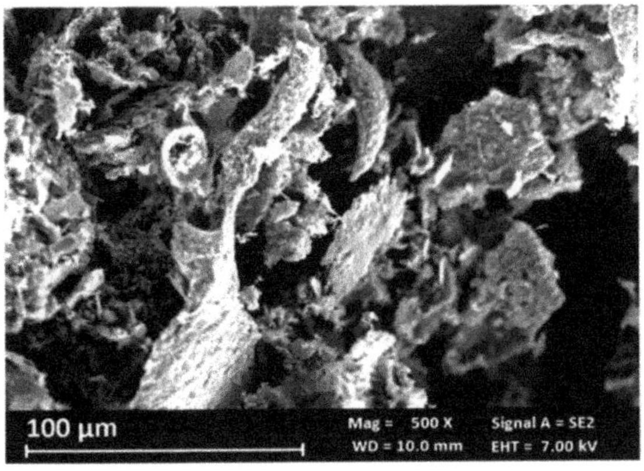

Figure 3. SEM images at 500× magnification: GE powder.

Table 2. EDX surface composition and EDX spectra of GE powder.

Atomic %	C	O	Al	Si
Graphene powder	95	3	traces	traces

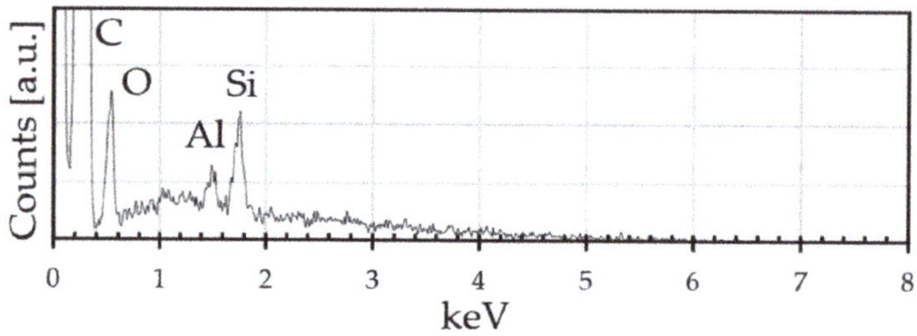

3.2. Potentiostatic

The morphology of the resulting graphene coatings was examined and characterized using optical and scanning electron microscopy (SEM), while energy-dispersive X-ray (EDX) analysis was employed to assess the elemental composition of the coatings (Table 3).

Table 3. EDX surface composition of GO deposit, CV deposit, and graphene powder.

Atomic %	C	O	Fe	S	Na	Al	Si
GO deposit	50	50	traces	/	/	/	/
CV deposit	30	41	26	1	2	/	/
Graphene powder	95	3	/	/	/	traces	traces

The deposits obtained at the end of polarization and the graph that shows the trend of the current during deposition are shown in Figure 4. The GO solution consistently results in the formation of good deposits, whereas the GE solution, conversely, rarely deposits, and when it does, the deposit is typically insignificant and dissolves rapidly.

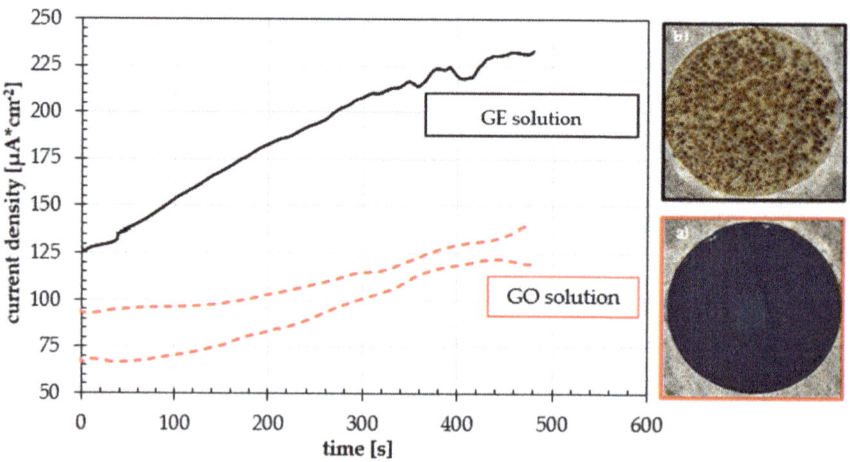

Figure 4. Current flowing between the test piece and the counter electrode in the GO solution and in the GE solution, and the surface of the specimens after a potentiostatic polarization of 10 V for 8 min in different solutions: (**a**) GO solution, and (**b**) GE solution.

Anodic currents increase as time passes. The conductivity of the two solutions is different and is about 66 kΩ/cm for the GO solution and about 80 kΩ/cm for the GE solution. The deposition obtained with chemically oxidized GO is much more homogeneous and covers the metallic substrate better than that obtained from the solution of unreduced graphene. According to the theory [25–27,36], negatively charged GO was deposited on the anode, while deposition was not observed on cathodes. In the absence of voltage, no visible film was formed on all electrodes.

From the digital microscope observation also shown in the figure, it can be observed that the deposit obtained with the GO solution has a high degree of coating and appears homogeneous, while the deposit obtained from the GE solution is inhomogeneous and has a very low degree of coating. Since the deposit obtained from the GE solution is very poor while that obtained from the GO solution is much better, a method of oxidizing the graphene contained in the GE solution by electrochemical means in order to create a solution of oxidized graphene more like the GO solution was sought.

Pristine graphene does not deposit through a constant-potential electrochemical process due to its unique structure and lack of reactive functional groups. Graphene oxide, however, due to its oxygen-containing functional groups, is more responsive to electrochemical methods.

3.3. Cyclic Voltammetry

Accordingly to the previous statement, graphene was deposited under variable potential tests, employing cyclic voltammetry. This approach involves cycling the applied potential over a defined range, promoting periodic adsorption and reduction processes at the electrode interface. This method thus provides a more suitable pathway for graphene deposition, overcoming the limitations associated with constant potential methods.

Cyclic voltammetry between 0 and +6 V was made in the GE solution (non-oxidized graphene solution) to study the evolution of the voltammetric curve. In order to be able to highlight the peaks related to graphene reactions, any other electrochemical contributions of the system must be excluded, i.e., the oxygen reaction evolution (OER) and the peaks of oxidation and growth of iron oxide in alkaline solution. For this reason, voltammetric curves obtained on the same material in the presence and absence of graphene in solution at the same pH were compared.

In Figure 5, the comparison between the first and last cycle of the series of voltammetries for the GE solution (blue) and for NaOH solution 0.1 M (orange) in the potential range between 0 and +6 V is shown.

In the range of water stability (up to about 0.6 V vs. SCE), the steel surface of the specimens is in a passive state [37], so the anodic current decreases as the number of voltammetric cycles increases. At high potentials, the curves become very disturbed, probably due to the oxygen bubbles effect. In graphene solution (GE solution), the anodic current is higher than that recorded in the NaOH solution alone and continues to increase after the onset of oxygen evolution, suggesting the involvement of a superimposed anodic process. During the anodic scan, there is a not well-defined broad peak around 4.4 V (Figure 6a): the peak current decreases with the increase in the number of voltammetry cycles, and the peak potential shifts to less anodic values. The areas corresponding to these peaks are highlighted with red stripes on the graph. The solution of NaOH 0.1 M does not present these peaks, confirming that it cannot be considered to represent electrochemical reactions involving iron ions only, since the same X65 specimens were used in both solutions and the oxidation of graphene.

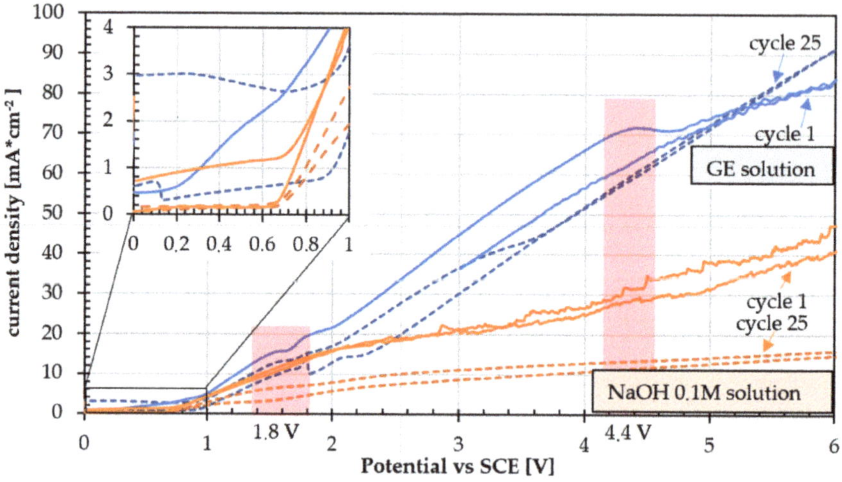

Figure 5. Comparison of series of voltammetry in graphene and NaOH 0.1 M solution from 0 to 6 V on API 5P X65 specimens.

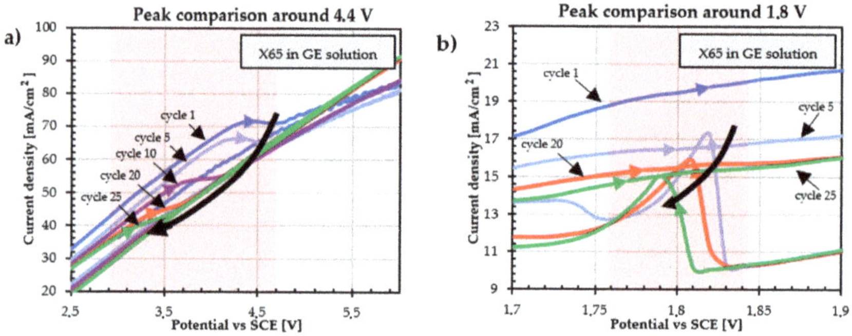

Figure 6. Comparison of GE solution peaks on API 5L X65 specimens: (**a**) around 4.4 V, and (**b**) around 1.8 V.

Given that graphene oxidation is nonstoichiometric, it is not feasible to ascertain an oxidation potential, but the presence of a black deposit on the metal surface suggests the growth of a carbon-rich coating on the substrate.

The presence of the small peaks in the return scan in the potential range around 1.8 V vs. SCE (Figure 6b) seems to indicate a partial reduction of the coatings, which could finally be formed by a mix of graphene and graphene oxide.

The possibility of having coatings on steel using cyclic voltammetry was further investigated by applying different voltammetry cycles to the specimens, varying the number of cycles and the potential range.

3.4. Effect of the Number of Cycles on the Coating

Depositions were made using 10, 25, and 100 cycles of voltammetry at a pH of 13. The 10-cycle tests were conducted to ensure a sufficient number of specimens were available for analysis within a reasonable timeframe (approximately 30 min per deposit). The 25-cycle and 100-cycle tests were conducted to evaluate the trend in deposit quality with respect to the number of voltammetric cycles. The observation of these deposits at SEM (Figure 7) shows the poor quality and low degree of coating of the 10-cycle deposits, whereas the 100-cycle deposits are thick and cover the whole surface but tend to fracture, a characteristic which is detrimental to a coating. The coatings obtained with 25 cycles are much more homogeneous than those obtained with 10 cycles, and there do not appear to be large cracks on the surface, as in the case of 100 cycles. Therefore, it was decided to continue the deposition by keeping the number of voltammetry cycles at 25.

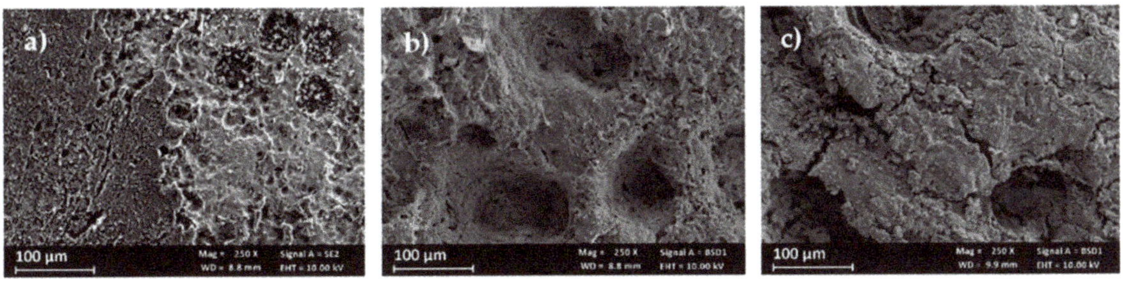

Figure 7. Surface of the specimen coated with CV at different numbers of cycles: (**a**) 10 cycles, (**b**) 25 cycles, and (**c**) 100 cycles.

Further confirmation of the better quality of the deposits obtained with 25 cycles compared to those obtained at 100 is provided by the analysis of their surface roughness using an interferometric microscope (Figure 8). The absorption coefficient of the coating is high because the graphene absorbs a lot of visible light, and this makes the process of optimizing the survey parameters more complex. The analyses were carried out on both the 100-cycle and 25-cycle specimens at different magnifications. From the analyses, Abbot diagrams were obtained, which show the distribution of the height of the reconstructed surface. The 25-cycle specimen histogram has a narrower base than the 100-cycle specimen, indicating that the average surface of the 25-cycle specimen is less uneven. Profiles from the surface analyzed reveal craters in both deposits.

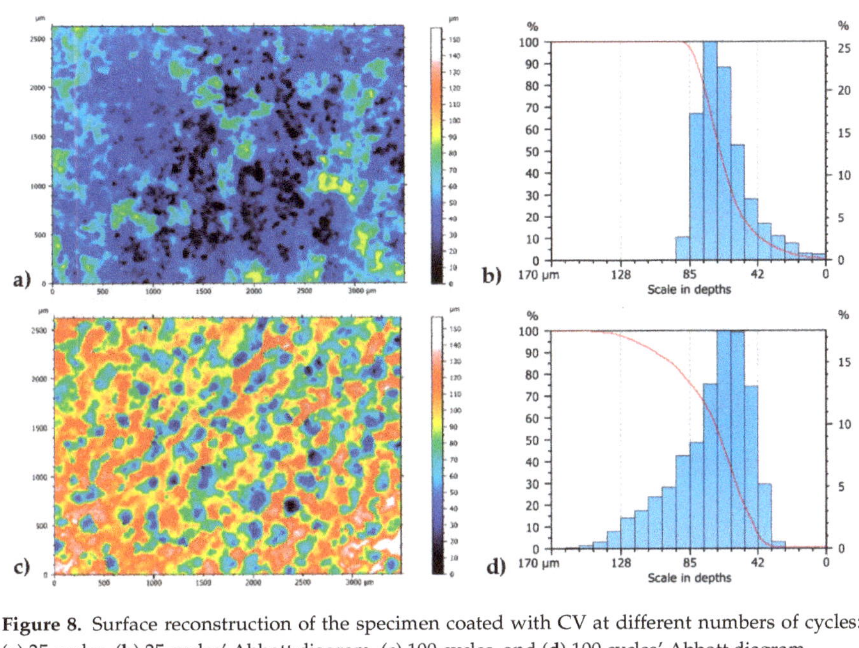

Figure 8. Surface reconstruction of the specimen coated with CV at different numbers of cycles: (**a**) 25 cycles, (**b**) 25 cycles' Abbott diagram, (**c**) 100 cycles, and (**d**) 100 cycles' Abbott diagram.

3.5. Effect of Potential Ranges

The graph in Figure 9 shows the potential ranges of the cyclic voltammetry carried out on the X65 specimens.

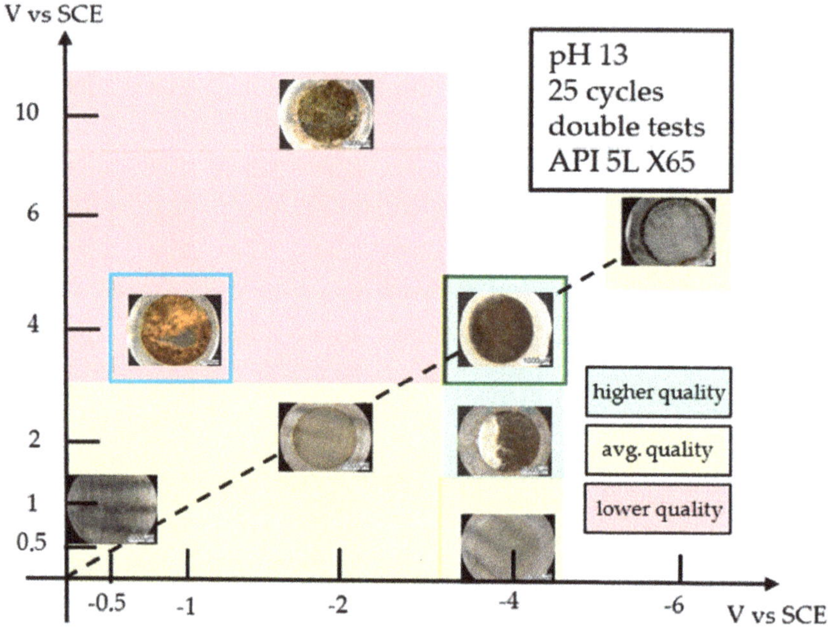

Figure 9. Surface areas of the deposits obtained as a function of the potential range scanned using CV with coating evaluation.

The deposit that appears to be most uniform and with the highest degree of coating is the one highlighted in the dark green box, obtained by scanning the potential range between −4 and +4 V (Figure 10a). Figure 11 shows the first, the tenth, and the last cyclic voltammetry cycles that produced this deposit.

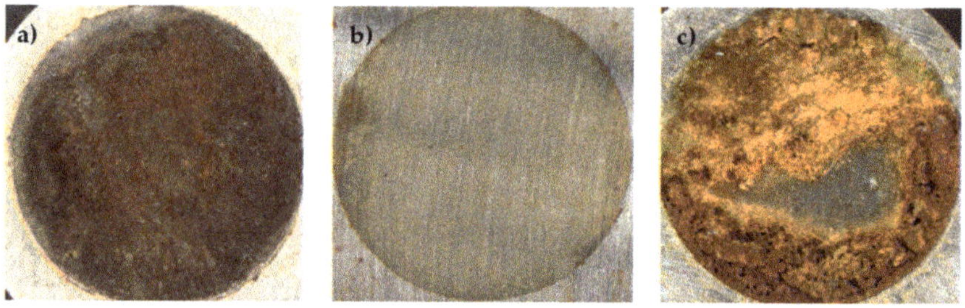

Figure 10. Surface of the deposits obtained after 25 voltammetry cycles at different potentials: (**a**) −4/+4 V, (**b**) −1/+4 V, and (**c**) −2/+2 V.

Deposits in the red area have an excessive amount of iron oxide that prevents the formation of a homogeneous graphene deposit. The deposits in the yellow area show the presence of a layer of graphene on the surface that appears homogeneously distributed over the whole surface, but its thickness is very thin and EDX surveys reveal the presence of large quantities of iron oxide and therefore it is not clear how much of this deposit is composed of graphene (Figure 10b). Deposits in the green area are those with consistent deposit thickness and a good degree of coating. Special mention should be made of the deposition between −1 and +4 V (blue box) because it shows a uniform graphene deposit but only in a restricted area of the specimen, while the rest of the surface is covered with a thick layer of iron oxide (Figure 10c).

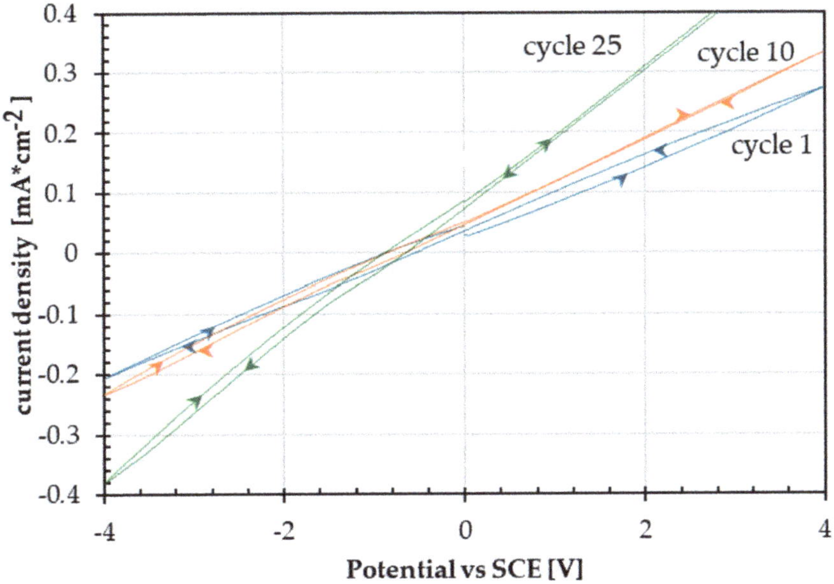

Figure 11. Comparison of first and last cyclic voltammetry cycles in GE solution from −4 to +4 V on API 5P X65.

3.6. Coating Characterization

Using as a reference the deposit obtained with 25 cycles of voltammetry between −4 and +4 V, analyses were made to investigate its chemical composition. From now on, this deposit will be referred to as a CV deposit, and the deposit obtained from the potentiostatic applied to the GO solution will be reported as the GO deposit.

SEM observations of these deposits show a clear difference in coating quality. GO deposit appears visually more homogeneous and less wrinkled, while the surface of the CV deposit is strewn with craters (Figure 12). Their thickness was observed by sectioning coated specimens embedded in resin. The GO deposit has an average thickness of about 10 microns, and the coating seems uniform and completely covers the observed surface. The CV deposit, on the other hand, has some areas with a thickness of around 5 microns, and other areas have a thin coating or probably no coating at all. The deterioration of the steel specimen surface due to the anodic polarization of the specimen during cyclic voltammetry can be observed when the steel specimen is looked at in sections. Nevertheless, this polarization is imperative for the oxidation of the oxidized graphene. The deposits analyzed in this study were obtained through polarization in an aqueous solution. It is probable that polarization in other environments might reduce the corrosion rate of the metal substrate. Tests to this effect will be the subject of a future paper.

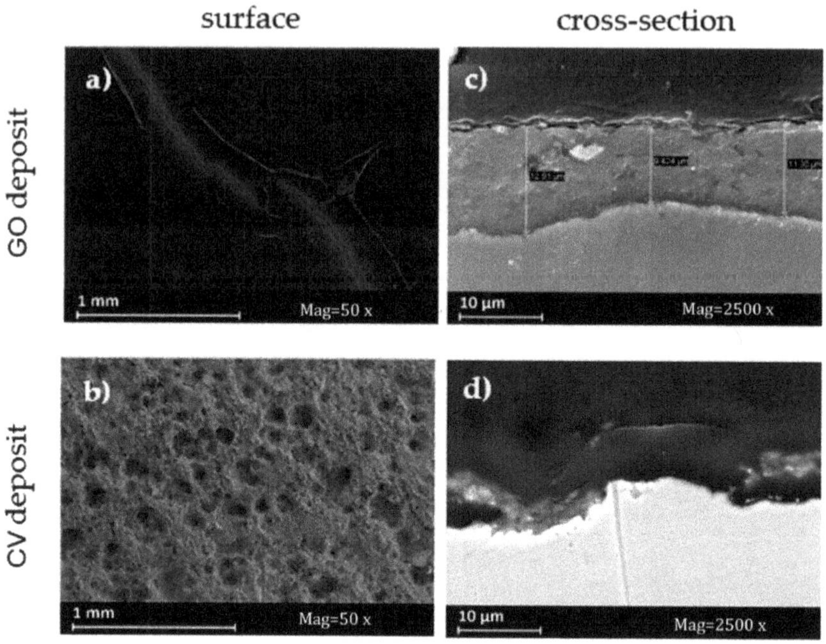

Figure 12. SEM images of the surface and cross section of (**a**,**c**) GO deposit and (**b**,**d**) CV deposit.

The EDX analysis (Figure 13) shows a difference between the two deposits. The GO deposit is 50% carbon and 50% oxygen, while the CV deposit is a combination of carbon, iron, and oxygen. The presence of carbon indicates the formation of a graphene-based deposit on the metal substrate, the presence of iron and oxygen indicates the formation of iron oxide and oxidized graphene, while sodium and sulfur are detected due to the addition of sodium laurylsulphate in solution. EDX compositions are shown in Table 3.

The deposits were also analyzed using Raman spectroscopy. This technique was used to establish whether the structures of both graphene and graphene oxide were somehow altered during the two respective selected electrochemical deposition processes. As a reference, both the GO and GE solutions were characterized by drop casting. Also, pristine GE powder was analyzed (Figure 14 and Table 4). As can be seen, the spectra of both the GO solution and GO deposit after potentiostatic polarization present practically the same peaks (D-, G-, 2D, D + G- and 2G- bands), with the same shape and position. Moreover, the ratio between the intensities of D-band respect to G-band (I_D/I_G), useful to evaluate the degree of order of the material (the lower it is, the closer is the sample to a graphene-like arrangement– sp^2 carbon; the higher it is, the more is the amount of sp^3 hybridized carbon), results only slightly reduced. This could be ascribed to a partial modification of the graphene oxide structure, upon potentiostatic deposition, becoming, most likely, moderately amorphous [38]. On the other hand, when GE powder is compared with GE solution, some evident differences arise. In particular, the position of D-band and G-band resulted in both shifting to higher values, meaning that the graphene layer reduced its thickness upon dispersion in water/ethanol solution [39], with an increased degree of disorder (in this case, I_D/I_G has passed from 0.65 to 0.79). The most evident change the GE material had experienced was due to the cyclic voltammetry deposition. With this method, the Raman spectra (CV deposit) revealed an amorphous material with a certain amount of sp^2 carbon kept upon deposition [40–45], which guarantees the electrical conductivity when graphitic material with mixed sp^2–sp^3 hybridization is used [42]. This analysis corroborates the hypothesis that during cyclic voltammetry, graphene is oxidized to GO, which is then deposited on the surface of the metal specimen. This deposit is different from that obtained by the potentiostatic deposition of pure GO, because during the cathodic scan of cyclic voltammetry, a partial reduction of GO can take place, forming the mix sp^2-sp^3 amorphous deposit, like the one obtained using the potentiostatic deposition of GO.

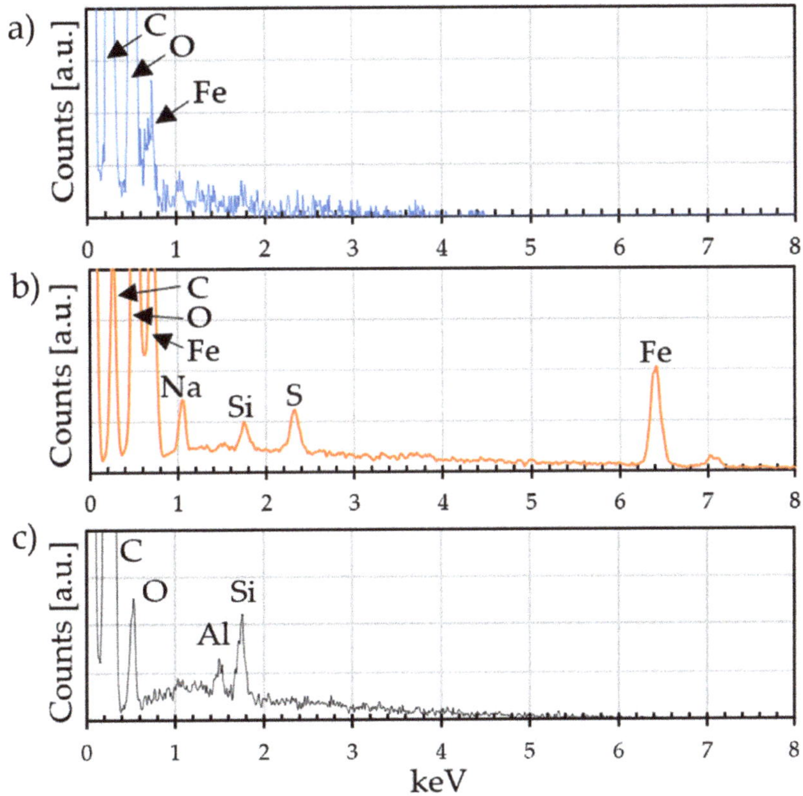

Figure 13. EDX spectra of surface composition of (**a**) GO deposit, (**b**) CV deposit, and (**c**) graphene powder.

Table 4. Raman results of the analyzed samples.

Sample	D-Position	G-Position	I_D/I_G
GO solution	1349.61	1601.83	1.30
GO deposit	1349.14	1601.43	1.22
GE powder	1336.83	1565.71	0.65
GE solution	1355.71	1589.99	0.79
CV deposit	1319.75	1571.78	0.71

The influence of the chemical composition of CV deposits (in particular, the presence of electrochemically produced iron oxide) on their ability to decrease hydrogen permeation should have been evaluated using tests according to the Devanathan–Stachurski method but both coatings suffer from cathodic disbonding. For this reason, other investigation techniques are being evaluated.

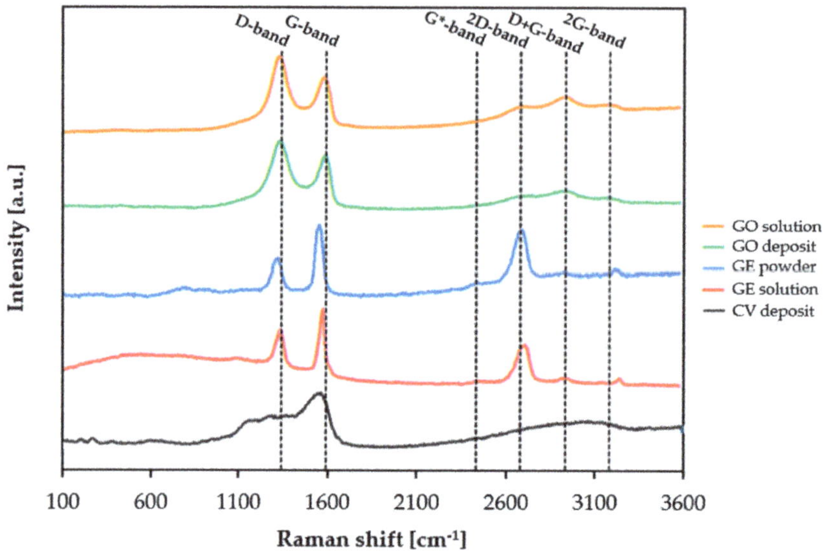

Figure 14. Raman spectra of GO solution, GO deposit, Ge powder, GE solution, and CV deposit (532 nm laser).

4. Conclusions

The aim of this study was to bypass the conventional step of converting graphite into graphene oxide (GO) through chemical oxidation and to utilize an electrochemical technique that enables in situ oxidation and subsequent rapid reduction directly on the metal surface. The principal findings of this study are:

- The potentiostatic deposition technique allows for the formation of uniform, smooth graphene oxide (GO) coatings with poor adhesion. However, this method is not suitable for the deposition of non-oxidized graphene.
- The deposition of graphene on a metal surface by direct electrochemical in situ oxidation and subsequent reduction via cyclic voltammetry represents a novel technique that has yet to be documented in the existing literature.
- Raman analyses have confirmed the presence of graphene, which was oxidized prior to reduction by cyclic voltammetry and deposited at the surface.
- SEM, EDX, and interferometric observations of the deposits show discontinuities, both in terms of morphology and composition, that can be attributed to the deterioration of the substrate, which has resulted in a reduction in the quality of the coating.

Although the deposits obtained with graphene oxidized using the cyclic voltammetry method do not yet have the same quality as those obtained using chemically oxidized graphene, this technique seems to have great potential both from the point of view of resources and time required to oxidize graphene and from the point of view of environmental impact. In the future, the cyclic voltammetry parameters will be optimized to improve the quality of the deposits, and once a homogeneous and uniform deposit is obtained, permeation tests will be carried out using the Devanathan–Stachurski method to evaluate the barrier effect of these coatings. During these tests, the problem of cathodic disbonding will have to be considered. To complete this study, this method will also be used to compare and evaluate the effectiveness of the barrier effect produced by the three types of coatings mentioned in this work (graphene DLC, GO deposit, and CV deposit). This will allow the evaluation of which factors could play a role in the quality and imperviousness of the coatings (e.g., high presence of iron oxide) and which can be optimized to improve the final characteristics of the coatings.

A critical consideration in the implementation of graphene coatings for the protection of hydrogen-exposed components concerns the potential release of chemisorbed gas. As demonstrated in the existing literature, the hydrogenation of graphene—through the formation of covalent C-H bonds—is a reversible process that can be activated thermally or by other energetic stimuli. As demonstrated by Elias et al. [46], hydrogen can be desorbed from hydrogenated graphene (graphane) through thermal treatment, effectively restoring the original graphene structure. However, the energy required to break the covalent C–H bond in graphene is estimated to be in the range of 2.5–3.0 eV per hydrogen atom. According to the Arrhenius equation, this energy corresponds to desorption temperatures in the range of approximately 300–500 °C [47,48]. Simulations conducted by Zhou et al. [49] further support this interpretation, indicating that at 300 °C the C–H bond remains thermally stable under standard conditions. As a result of the above considerations, we might assume that the release of hydrogen accumulated in the graphene coating under ambient conditions is unlikely. On the other hand, in confined environments such as pipelines or reservoirs, such a release, although probably negligible, could lead to a danger of H_2 accumulation in explosive concentrations during maintenance operations as a result of oxygen entering the pipeline and this possibility suggests the need to consider the dynamic behavior of adsorbed hydrogen on graphene coatings in future studies and in the engineering of graphene coatings for applications in hydrogen-containing environments.

Author Contributions: Conceptualization, M.C. and B.A.; methodology, M.C.; validation, M.C., B.A. and G.C.L.; formal analysis, M.P.; investigation, M.P. and R.P.R.; resources, B.A.; data curation, M.P.; writing—original draft preparation, M.P.; writing—review and editing, M.C., B.A. and G.C.L.; supervision, M.C., B.A. and G.C.L.; project administration, B.A.; funding acquisition, B.A. All authors have read and agreed to the published version of the manuscript.

Funding: This research was funded by the European Union—NextGeneration EU from the Italian Ministry of Environment and Energy Security POR H2 AdP MMES/ENEA with involvement of CNR and RSE, PNRR—Mission 2, Component 2; Investment 3.5 "Ricerca e sviluppo sull'idrogeno", CUP: I83C22001170006.

Institutional Review Board Statement: Not applicable.

Informed Consent Statement: Not applicable.

Data Availability Statement: The raw data supporting the conclusions of this article will be made available by the authors on request.

Conflicts of Interest: The authors declare no conflicts of interest.

References

1. Capros, P.; Paroussos, L.; Fragkos, P.; Tsani, S.; Boitier, B.; Wagner, F.; Busch, S.; Resch, G.; Blesl, M.; Bollen, J. Description of models and scenarios used to assess European decarbonisation pathways. *Energy Strat. Rev.* **2014**, *2*, 220–230. [CrossRef]
2. Bennoua, S.; Le Duigou, A.; Quéméré, M.-M.; Dautremont, S. Role of hydrogen in resolving electricity grid issues. *Int. J. Hydrogen Energy* **2015**, *40*, 7231–7245. [CrossRef]
3. Chamandoust, H.; Hashemi, A.; Bahramara, S. Energy management of a smart autonomous electrical grid with a hydrogen storage system. *Int. J. Hydrogen Energy* **2021**, *46*, 17608–17626. [CrossRef]
4. Available online: https://publications.jrc.ec.europa.eu/repository/handle/JRC126763 (accessed on 10 September 2024).
5. Gangloff, R.P.; Somerday, B.P. (Eds.) The problem, its characterisation and effects on particular alloy classes. In *Gaseous Hydrogen Embrittlement of Materials in Energy Technologies*; Woodhead Publishing Limited: Cambridge, UK, 2012; Volume 1.
6. Lei, Y.; Hosseini, E.; Liu, L.; Scholes, C.A.; Kentish, S.E. Internal polymeric coating materials for preventing pipeline hydrogen embrittlement and a theoretical model of hydrogen diffusion through coated stee. *Int. J. Hydrogen Energy* **2022**, *47*, 31409–31419. [CrossRef]

7. Shen, S.; Song, X.; Li, Q.; Li, X.; Zhu, R.; Yang, G. Effect of CrxCy–NiCr coating on the hydrogen embrittlement of 17-4 PH stainless steel using the smooth bar tensile test. *J. Mater. Sci.* **2019**, *54*, 7356–7368. [CrossRef]
8. Morimoto, T.; Kumai, T. Prevention of Hydrogen Embrittlement Using Ultra Rapid Cooling Thermal Spraying Gun. *ISIJ Int.* **2017**, *57*, 1461–1467. [CrossRef]
9. Bhadeshia, H.K.D.H. Prevention of Hydrogen Embrittlement in Steels. *ISIJ Int.* **2016**, *56*, 24–36. [CrossRef]
10. Khare, A.; Vishwakarma, M.; Ahmed, S. Combating Hydrogen Embrittlement With Graphene Based Coatings. *Int. J. Adv. Res. Eng. Technol.* **2019**, *10*, 234–251. [CrossRef]
11. Coatings and Liners for Hydrogen Service Pipelines. Available online: https://www.csagroup.org/ (accessed on 10 September 2024).
12. Wan, H.; Cheng, Z.l.; Song, D.; Chen, C. Preparation and performance study of waterborne epoxy resin/non-covalent modified graphene oxide hydrogen barrier coatings. *Int. J. Hydrogen Energy* **2024**, *53*, 218–228. [CrossRef]
13. Cai, K.; Jiang, B. Preparation and Characterization of Composite Hydrogen Barrier Coatings with (Graphene–Epoxy Resin)/(Silicon Carbide–Epoxy Resin)/(Graphene–Epoxy Resin) Sandwich Structures. *Coatings* **2025**, *15*, 518. [CrossRef]
14. Soldano, C.; Mahmood, A.; Dujardin, E. Production, properties and potential of graphene. *Carbon* **2010**, *48*, 2127–2150. [CrossRef]
15. Prasai, D.; Tuberquia, J.C.; Harl, R.R.; Jennings, G.K.; Bolotin, K.I. Graphene: Corrosion-Inhibiting Coating. *ACS Nano* **2012**, *6*, 1102–1108. [CrossRef] [PubMed]
16. Berry, V. Impermeability of graphene and its applications. *Carbon* **2013**, *62*, 1–10. [CrossRef]
17. Bunch, J.S.; Verbridge, S.S.; Alden, J.S.; Van Der Zande, A.M.; Parpia, J.M.; Craighead, H.G.; McEuen, P.L. Impermeable atomic membranes from graphene sheets. *Nano Lett.* **2008**, *8*, 2458–2462. [CrossRef]
18. Sun, P.Z.; Yang, Q.; Kuang, W.J.; Stebunov, Y.V.; Xiong, W.Q.; Yu, J.; Nair, R.R.; Katsnelson, M.I.; Yuan, S.J.; Grigorieva, I.V.; et al. Limits on gas impermeability of graphene. *Nature* **2020**, *579*, 229–232. [CrossRef]
19. Nam, T.H.; Lee, J.H.; Choi, S.R.; Yoo, J.B.; Kim, J.G. Graphene coating as a protective barrier against hydrogen embrittlement. *Int. J. Hydrogen Energy* **2014**, *39*, 11810–11817. [CrossRef]
20. Shi, K.; Xiao, S.; Ruan, Q.; Wu, H.; Chen, G.; Zhou, C.; Jiang, S.; Xi, K.; He, M.; Chu, P.K. Hydrogen permeation behavior and mechanism of multi-layered graphene coatings and mitigation of hydrogen embrittlement of pipe steel. *Appl. Surf. Sci.* **2022**, *573*, 151529. [CrossRef]
21. Miglietta, M.L.; Polichetti, T.; Massera, E.; Mauro, A.D.G.D.; Alfano, B. Direct Growth Of Graphene On Stainless Steel As Protective Layer Against Hydrogen Corrosion. In Proceedings of the 2023 IEEE Nanotechnology Materials and Devices Conference (NMDC), Paestum, Italy, 22–25 October 2023; pp. 257–258. [CrossRef]
22. Saeed, M.; Alshammari, Y.; Majeed, S.A.; Al-Nasrallah, E. Chemical Vapour Deposition of Graphene—Synthesis, Characterisation, and Applications: A Review. *Molecules* **2020**, *25*, 3856. [CrossRef]
23. Li, X.; Cai, W.; An, J.; Kim, S.; Nah, J.; Yang, D.; Piner, R.; Velamakanni, A.; Jung, I.; Tutuc, E.; et al. Large-Area Synthesis of High-Quality and Uniform Graphene Films on Copper Foils. *Science* **2009**, *324*, 1312–1314. [CrossRef]
24. Yu, H.; Zhang, B.; Bulin, C.; Li, R.; Xing, R. High-efficient Synthesis of Graphene Oxide Based on Improved Hummers Method. *Sci. Rep.* **2016**, *6*, 36143. [CrossRef]
25. Jabbar, A.; Yasin, G.; Khan, W.Q.; Anwar, M.Y.; Korai, R.M.; Nizam, M.N.; Muhyodin, G. Electrochemical deposition of nickel graphene composite coatings: Effect of deposition temperature on its surface morphology and corrosion resistance. *RSC Adv.* **2017**, *7*, 31100–31109. [CrossRef]
26. Liu, C.; Wei, D.; Huang, X.; Mai, Y.; Zhang, L.; Jie, X. Electrodeposition of Co–Ni–P/graphene oxide composite coating with enhanced wear and corrosion resistance. *J. Mater. Res.* **2019**, *34*, 1726–1733. [CrossRef]
27. Tseluikin, V.; Dzhumieva, A.; Yakovlev, A.; Mostovoy, A.; Lopukhova, M. Electrodeposition of Graphene Oxide Modified Composite Coatings Based on Nickel-Chromium Alloy. *Crystals* **2021**, *11*, 415. [CrossRef]
28. Kim, Y.-S.; Kim, J.-G. Electroplating of reduced-graphene oxide on austenitic stainless steel to prevent hydrogen embrittlement. *Int. J. Hydrogen Energy* **2017**, *42*, 27428–27437. [CrossRef]
29. Behunová, D.M.; Gallios, G.; Girman, V.; Kolev, H.; Kaňuchová, M.; Dolinská, S.; Václavíková, M. Electrophoretic Deposition of Graphene Oxide on Stainless Steel Substrate. *Nanomaterials* **2021**, *11*, 1779. [CrossRef]
30. Yu, Q.; Wei, L.; Yang, X.; Wang, C.; Chen, J.; Du, H.; Shen, W.; Kang, F.; Huang, Z.-H. Electrochemical synthesis of graphene oxide from graphite flakes exfoliated at room temperature. *Appl. Surf. Sci.* **2022**, *598*, 153788. [CrossRef]
31. Lama, G.C.; Santillo, C.; Recupido, F.; Liu, J.; Verdolotti, L.; Marzella, R.; Polichetti, T.; Kaciulis, S.; Lavorgna, M. Autoclave-mediated reduction of graphene oxide for enhanced conductive films. *Appl. Surf. Sci.* **2024**, *657*, 159741. [CrossRef]
32. Alfano, B.; Polichetti, T.; Miglietta, M.L.; Massera, E.; Schiattarella, C.; Ricciardella, F.; Di Francia, G. Fully eco-friendly H_2 sensing device based on Pd-decorated graphene. *Sens. Actuators B Chem.* **2017**, *239*, 1144–1152. [CrossRef]
33. *ASTM G5-14*; Standard Reference Test Method for Making Potentiodynamic Anodic Polarization Measurements. ASTM International: West Conshohocken, PA, USA, 2021. [CrossRef]

34. Elgrishi, N.; Rountree, K.J.; McCarthy, B.D.; Rountree, E.S.; Eisenhart, T.T.; Dempsey, J.L. Practical Beginner's Guide to Cyclic Voltammetry. *J. Chem. Educ.* **2018**, *95*, 197–206. [CrossRef]
35. Aoki, K.; Kato, N. Analysis of the cyclic voltammograms associated with deposition or precipitation of the electrochemical product. *J. Electroanal. Chem. Interfacial Electrochem.* **1988**, *245*, 51–60. [CrossRef]
36. Mbayachi, V.B.; Ndayiragije, E.; Sammani, T.; Taj, S.; Mbuta, E.R.; Khan, A.U. Graphene synthesis, characterization and its applications: A review. *Results Chem.* **2021**, *3*, 100163. [CrossRef]
37. Cabrini, M.; Lorenzi, S.; Pastore, T. Cyclic voltammetry evaluation of inhibitors for localised corrosion in alkaline solutions. *Electrochim. Acta* **2014**, *124*, 156–164. [CrossRef]
38. de Rességuier, T.; Kurakevych, O.O.; Chabot, A.; Petitet, J.P.; Solozhenko, V.L. Structural changes and phase stability of graphitelike BC3 under explosive shock-wave loading. *J. Appl. Phys.* **2010**, *108*, 083522. [CrossRef]
39. Wall, M. Raman Spectroscopy Optimizes Graphene Characterization. *AM&P Tech. Artic.* **2012**, *170*, 35–38.
40. Ferralis, N. Probing mechanical properties of graphene with Raman spectroscopy. *J. Mater. Sci.* **2010**, *45*, 5135–5149. [CrossRef]
41. Saito, R.; Hofmann, M.; Dresselhaus, G.; Jorio, A.; Dresselhaus, M.S. Raman spectroscopy of graphene and carbon nanotubes. *Adv. Phys.* **2011**, *60*, 413–550. [CrossRef]
42. Vejpravová, J. Mixed sp2–sp3 nanocarbon materials: A status quo review. *Nanomaterials* **2021**, *11*, 2469. [CrossRef] [PubMed]
43. Brzhezinskaya, M.; Irzhak, A.; Irzhak, D.; Kang, T.W.; Kononenko, O.; Matveev, V.; Panin, G.; Roshchupkin, D. Direct growth of graphene film on piezoelectric $La_3Ga_{5.5}Ta_{0.5}O_{14}$ crystal. *Rapid Res. Lett.* **2016**, *10*, 639–644. [CrossRef]
44. Brzhezinskaya, M.; Kapitanova, O.O.; Kononenko, O.V.; Koveshnikov, S.; Korepanov, V.; Roshchupkin, D. Large-scalable graphene oxide films with resistive switching for non-volatile memory applications. *J. Alloys Compd.* **2020**, *849*, 156699. [CrossRef]
45. Kononenko, O.; Brzhezinskaya, M.; Zotov, A.; Korepanov, V.; Levashov, V.; Matveev, V.; Roshchupkin, D. Influence of numerous Moiré superlattices on transport properties of twisted multilayer graphene. *Carbon* **2022**, *194*, 52–61. [CrossRef]
46. Elias, D.C.; Nair, R.R.; Mohiuddin, T.M.G.; Morozov, S.V.; Blake, P.; Halsall, M.P.; Ferrari, A.C.; Boukhvalov, D.W.; Katsnelson, M.I.; Geim, A.K.; et al. Control of Graphene's Properties by Reversible Hydrogenation: Evidence for Graphane. *Science* **2009**, *323*, 610–613. [CrossRef] [PubMed]
47. Delfino, F.; Ros, C.; Palardonio, S.M.; Carretero, N.M.; Murcia-López, S.; Morante, J.R.; Martorell, J.; Fthenakis, Z.G.; Sgroi, M.F.; Tozzini, V.; et al. Multi-methodological analysis of hydrogen desorption from graphene. *Carbon* **2024**, *227*, 119211. [CrossRef]
48. Openov, L.A.; Podlivaev, A.I. Thermal desorption of hydrogen from graphane. *Tech. Phys. Lett.* **2010**, *36*, 31–33. [CrossRef]
49. Zhou, J.; Wang, Q.; Sun, Q.; Chen, X.S.; Kawazoe, Y.; Jena, P. Ferromagnetism in Semihydrogenated Graphene Sheet. *Nano Lett.* **2009**, *9*, 3867–3870. [CrossRef] [PubMed]

materials

MDPI

Article

Effect of Al on the Oxidation Behavior of TiCrZrNbTa High-Entropy Coatings on Zr Alloy

Min Guo [1,2], Chaoyang Chen [1,2], Bin Song [1,2], Junhong Guo [1,2], Junhua Hu [1,2] and Guoqin Cao [1,2,*]

[1] Lonemen Laboratory, Luoyang 471000, China; g19960413mbs@163.com (M.G.); 13283071567@163.com (C.C.); songbinzzu@163.com (B.S.); 15835436647@163.com (J.G.); 15238209837@163.com (J.H.)

[2] School of Materials Science and Engineering, Zhengzhou University, Zhengzhou 450001, China

* Correspondence: caoguoqin@zzu.edu.cn

Abstract: This study investigates the role of Al alloying in tailoring the oxidation resistance of AlTiCrZrNbTa refractory high-entropy alloy (RHEA) coatings on Zry-4 substrates under high-temperature steam environments. Coatings with varying Al contents (0–25 at.%) were deposited via magnetron sputtering and subjected to oxidation tests at 1000–1100 °C. The results demonstrate that Al content critically governs oxidation kinetics and coating integrity. The optimal performance was achieved at 10 at.% Al, above which a dense, continuous composite oxide layer (Al_2O_3, TiO_2, Cr_2O_3) formed, effectively suppressing oxygen penetration and maintaining strong interfacial adhesion. Indentation tests confirmed enhanced mechanical integrity in Al-10 coatings, with minimal cracking post-oxidation. Excessive Al alloying ($\geq$17 at.%) led to accelerated coating oxidation. This work establishes a critical Al threshold for balancing oxidation and interfacial bonding, providing a design strategy for developing accident-tolerant fuel cladding coatings.

Keywords: refractory high-entropy alloy coatings; aluminum alloying; protective oxide layer; interfacial bonding; accident-tolerant fuel cladding

1. Introduction

Due to the Fukushima nuclear accident, public concerns about nuclear safety have increased, significantly impacting the global nuclear power industry [1–3]. In response, the international community has been dedicated to developing a new type of nuclear fuel capable of withstanding similar accidents, known as accident-tolerant fuel (ATF) [4,5]. One potential approach is the application of protective ATF coatings on the surface of fuel cladding [6]. Various candidate coating materials and deposition techniques have been explored for zirconium alloy fuel cladding. Current research on fuel cladding coatings mainly includes metal coatings [7–10] and ceramic coatings [11–15]. Tang et al. [14] compared the performance of different coatings under both normal operating conditions (subcritical water) and accident conditions (high-temperature steam). Al- and Si-rich coatings exhibited promising behavior under accident conditions but performed poorly under normal conditions. Ti- and Ni-based coatings failed to adapt to high-temperature steam, whereas Cr-based coatings demonstrated superior performance under both conditions, making them a promising choice for cladding materials. However, at elevated temperatures, elemental interdiffusion between Cr coatings and Zr alloy substrates leads to premature coating failure. Thus, further optimization of coating compositions is necessary to enhance their resistance to extreme conditions.

High-entropy alloys (HEAs) exhibit unique properties, including high hardness, wear resistance, corrosion resistance, and thermal stability, due to high-entropy effects, lattice dis-

https://doi.org/10.3390/ma18091997

tortion, sluggish diffusion, and the "cocktail effect" [16–19]. These properties make HEAs highly promising for surface protection applications. To ensure suitability for extreme environments such as aerospace and nuclear reactors, new alloys should be composed of high-melting-point elements. Based on HEA concepts, Senkov et al. [20] introduced the concept of refractory high-entropy alloys (RHEAs), which differ from traditional HEAs by incorporating refractory metals instead of transition metals. RHEAs exhibit excellent mechanical properties [21–23] and radiation resistance [24–27]. However, due to the high oxygen affinity of refractory metals, severe oxidation occurs at elevated temperatures [28–31]. To counteract this, researchers have explored the addition of elements such as Al and Si to improve the oxidation resistance. Al_2O_3 layers can reduce oxidation kinetics, while Si prevents the formation of volatile vanadium oxides and promotes the development of a stable oxide layer while minimizing nitridation [28–31]. Anber et al. [29] investigated the effect of Al addition (4.8 and 13 at.%) on HfNbTaTiZr alloys and conducted oxidation tests at 900 °C. The results showed that Al significantly reduced the oxidation weight gain rate. Similarly, Pei et al. [30] studied the impact of Al on the oxidation resistance of $TiZrV_{0.5}Nb_{0.5}$ RHEAs, preparing $TiZrV_{0.5}Nb_{0.5}$, $TiZrV_{0.5}Nb_{0.5}Si_{0.3}$, and $TiZrV_{0.5}Nb_{0.5}Al_{0.7}$ alloys. Oxidation experiments at 1000 °C revealed that all samples exhibited improved oxidation resistance, with $TiZrV_{0.5}Nb_{0.5}Al_{0.7}$ forming an Al_2O_3 protective layer. The dense oxide structure inhibited extensive inward oxygen diffusion, thus preventing severe internal oxidation. The low oxygen diffusion rate also suppressed the formation of volatile oxides, thereby preventing oxide layer cracking.

Given the safety concerns associated with traditional zirconium alloys, there is an urgent need to develop high-performance coatings. This study combines the high-temperature oxidation resistance of Cr-based coatings with the superior properties of RHEAs. Ti, Cr, Zr, Nb, and Ta were selected as refractory metals to provide excellent mechanical properties and high-temperature stability, while Al was incorporated to enhance oxidation resistance. Using magnetron sputtering, TiCrZrNbTa coatings with varying Al contents were deposited onto Zry-4 alloy substrates. High-temperature steam oxidation tests were conducted at 1000–1100 °C to evaluate the effect of Al content on oxidation resistance. The optimal coating composition was identified, and the morphology, structure, and element diffusion behavior of the oxidized coatings were analyzed in detail.

2. Experimental Details

A magnetron sputtering system was used to deposit refractory high-entropy alloy (RHEA) coatings onto the zircaloy-4(Zry-4) substrates. The chemical composition of the Zry-4 is Zr-98.2%, Sn-1.5%, Fe-0.2%, and Cr-0.1%. A circular target with a diameter of 76.2 mm was composed of sector-shaped metallic targets of Al, Ti, Cr, Zr, Nb, and Ta, positioned at different angles. The sources of Zry-4 and metallic targets were made by (ZhongNuo Advanced Material Technology Co., Ltd., Beijing, China). The AlTiCrZrNbTa coatings were deposited onto polished Zry-4 alloy substrates using direct current (DC) magnetron sputtering. The composition of the coatings was adjusted by varying the proportions of target materials based on their respective sputtering yields, while the sputtering current and power were fine-tuned to control the final coating composition.

A tube furnace was preheated to the target temperatures (1000 °C and 1100 °C) prior to placing the coated samples inside for the isothermal experiment. A steam environment was established by using ultrapure water as the steam source. Steam was introduced at a flow rate of 0.2 mL/min after the furnace reached 150 °C. The high-temperature steam oxidation experiments were conducted by exposing the samples to 1000 °C for 60 min and 1100 °C for 30 min in the furnace, followed by air cooling. The thickness of the as-deposited coatings was measured using a Dektak XT profilometer (Shanghai R&D Instrument Co., Ltd.,

Shanghai, China). To minimize measurement error, each sample was measured three times, and the average value was taken as the final thickness.

Mechanical tests were conducted on the oxidized coatings. A microhardness tester equipped with a square-shaped indenter was used. The applied load was set to 0.98 N, with a dwell time of 15 s. The indentation size was in the micron range, allowing for an overall evaluation of the mechanical properties across the coating thickness. The coated samples were weighed before and after oxidation at various time intervals using an electronic balance. To reduce measurement error, each sample was weighed at least three times.

The macroscopic morphology of the coatings after high-temperature steam oxidation was characterized using an optical microscope (OM) from Zeiss, Oberkochen, Germany (Axio Scope A1 Pol). The cross-sectional samples were ground and polished sequentially using SiC sandpapers with grit sizes of 100–2000#, followed by fine polishing with W2.5 μm and W0.5 μm diamond suspensions. The phase composition of the coatings was analyzed using an X-ray diffraction (XRD) system (Empyrean, PANalytical, Almelo, The Netherlands) with a Cu target as the radiation source. Measurements were performed using the grazing incidence X-ray diffraction (GIXRD) mode with a parallel beam setup. A Helios G4 CX dual-beam scanning electron microscope (SEM) from Thermo Fisher Scientific (Brno, Czech Republic) was used to analyze the coating morphology. Both secondary electron (SE) and backscattered electron (BSE) modes were employed, with a resolution of $\leq$0.6 nm in SE mode and $\leq$2 nm in BSE mode at high voltage. Additionally, an energy-dispersive spectroscopy (EDS) system was integrated into the SEM to analyze elemental distribution and composition, providing insights into the oxidation resistance of the coatings.

3. Results and Discussion

AlTiCrZrNbTa coatings with varying Al content were deposited on Zry-4 alloy substrates using single-target magnetron sputtering. The compositions of the as-sputtered coatings are listed in Table 1. Based on Al content, the coatings were designated as Al-0, Al-5, Al-10, Al-17, and Al-25. All coatings maintained a consistent thickness of ~4 μm to eliminate thickness effects on oxidation behavior. The ideal mixing entropy was calculated by the following equation [32]:

$$\Delta S_{mix} = -R\sum x_i \ln x_i$$

where R is the gas constant and x_i represents the molar fraction of element i. Thermodynamic calculations revealed that increasing Al content enhanced the ideal mixing entropy from 1.46R to 1.73R, aligning with high-entropy alloy design principles.

Table 1. Composition and mixing entropy of AlTiCrZrNbTa coatings.

Sample	Element (at.%)						ΔS_{mix}
	Ti	Cr	Zr	Nb	Ta	Al	
Al-0	20.49	41.04	14.79	16.36	7.32	-	1.46R
Al-5	18.56	37.48	13.91	15.85	7.74	6.46	1.59R
Al-10	18.41	35.55	11.82	15.78	7.28	11.16	1.64R
Al-17	16.13	33.6	11.53	15.24	6.07	17.43	1.67R
Al-25	18.69	16.42	16.21	17.27	6.01	25.4	1.73R

Figure 1a–e display SEM surface images of the coatings, revealing dense and defect-free morphologies. XRD patterns in Figure 1f exhibit a broad diffraction peak, indicating an amorphous structure, consistent with previous reports on sputtered high-entropy alloy

coatings [33,34]. Notably, Al incorporation induced gradual peak shifts toward higher angles, suggesting a denser atomic packing density.

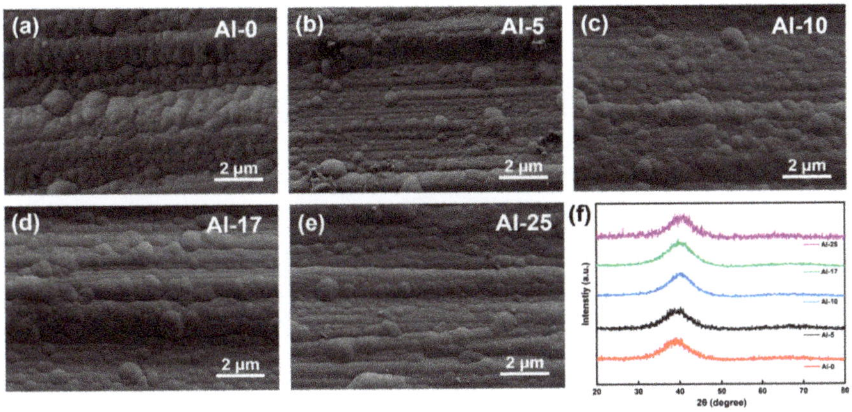

Figure 1. SEM images of the surfaces of AlTiCrZrNbTa coatings in sputtered states: (**a**) Al-0 coating; (**b**) Al-5 coating; (**c**) Al-10 coating; (**d**) Al-17 coating; (**e**) Al-25 coating; (**f**) XRD spectra of AlTiCrZrNbTa coatings in sputtered states.

To investigate the influence of Al content on the oxidation resistance of high-entropy alloy coatings against high-temperature steam, the oxidation experiment at 1000 °C was conducted on Zry-4 alloy with Al_xTiCrZrNbTa (x = 0, 5, 10, 17, 25 at.%) coatings. In Figure 2, it can be observed that the oxidation weight gain of all coating systems follows a parabolic pattern. The parabolic rate constant, k_p, was calculated by the mass gain per unit area of a sample ($\Delta W/A$) and oxidation time (t) according to Equation [35]:

$$\left(\frac{\Delta W}{A}\right)^2 = k_p t$$

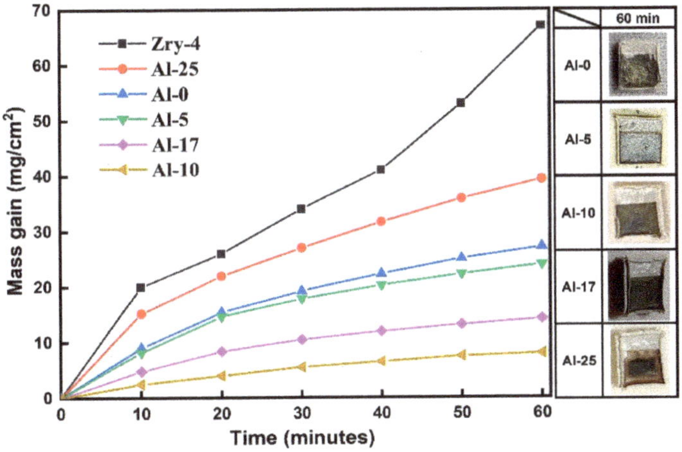

Figure 2. Specific mass change as a function of time for AlTiCrZrNbTa coatings during isothermal exposure to steam at 1000 °C and macroscopic morphology of the sample after oxidation.

The parabolic rate constant, k_p, is about 13.1, 10, 1.2, 3.5, 25.9 mg^2 cm^{-4} min^{-1} for Al-0, Al-5, Al-10, Al-17, Al-25, respectively. In addition, a comparison of the k_p values obtained in this work with literature data [35–37] indicates that the coatings fabricated in this study possess excellent resistance to high-temperature oxidation. The Al content significantly affects the oxidation kinetics behavior. Coatings with low Al content ($x \leq 5$) have a higher oxidation rate, and the substrate undergoes significant deformation accompanied by large-scale coating delamination; when the Al content increases to 10 at.%, the oxidation weight gain is the lowest, and the coating maintains high integrity, indicating its good oxygen-blocking ability, while coatings with excessively high Al content ($x \geq 17$) result in a further increase in oxidation weight gain, and the deformation of the substrate and coating delamination become more severe.

Figure 3 shows the OM and SEM images of the surface and cross-section morphology after coating oxidation. From the OM images (Figure 3a–e,a1–e1), it can be observed that a large number of cracks appeared on the surfaces of Al-0, Al-5, Al-17, and Al-25 coatings, which provided channels for O to directly contact and react with the Zr matrix. However, the surface of the Al-10 coating remained relatively smooth after oxidation, without obvious macroscopic defects, which preliminarily verified its superior oxidation resistance. The statistical results of the experimental data are shown in Table 2. Only the Al-10 sample remained intact after high-temperature water vapor oxidation, and the coating effectively blocked the invasion of the O element, while the Al-0, Al-5, Al-17, and Al-25 coatings failed after long-term oxidation. The preliminary judgment indicates that an appropriate amount of Al alloying can improve the oxidation resistance of the coating, but when the Al content exceeds 15 at.%, the coating's oxygen-blocking ability decreases.

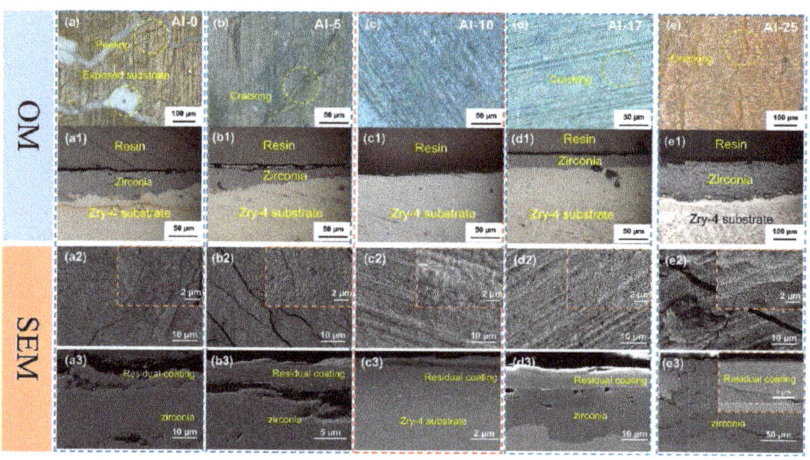

Figure 3. Surface and cross-sectional morphologies of Al-0, Al-5, Al-10, Al-17, and Al-25 coatings after oxidation at 1000 °C for 60 min: **(a–a3)** Al-0 coating; **(b–b3)** Al-5 coating; **(c–c3)** Al-10 coating; **(d–d3)** Al-17 coating; **(e–e3)** Al-25 coating.

Table 2. Protective evolution of AlTiCrZrNbTa coatings at 1000 °C for 60 min.

Sample	Temperature & Time
	1000 °C 60 min
Al-0	Failed
Al-5	Failed
Al-10	Effective
Al-17	Failed
Al-25	Failed

A dense and continuous oxide layer can effectively prevent O from eroding the residual coating and the substrate [31]. From the SEM morphology analysis, the TiCrZrNbTa coating without Al alloying formed granular oxides after oxidation (Figure 3a2,a3), and the cross-sectional morphology further shows that the coating cracked and was not continuous after oxidation, and was completely separated from the substrate. The high-magnification images showed that the granular oxides were more densely combined for the Al-5 coating. However, at the macroscopic scale, there were still a large number of cracks on the coating surface. The cross-sectional images (Figure 3b3) indicated that the coating remained relatively continuously distributed on the substrate, but was still separated from the substrate. With further increase in Al content, the Al-10 coating demonstrated superior oxidation, and the surface was mainly composed of dense granular oxides (Figure 3c2). The cross-sectional morphology (Figure 3c3) showed that the coating exhibited strong bonding with the substrate, and no obvious cracks or defects were observed at the interface. The oxidized surface remained smooth for the Al-17 coating (Figure 3d2), but there were holes in local areas. The cross-sectional morphology (Figure 3d3) showed that the coating was intact without obvious cracks or defects and was closely bonded to the substrate. However, the oxidation of the Zr substrate was identified, indicating that the coating failed to form an effective oxygen-blocking structure. When the Al content further increased to 25 at.%, cracks reappeared on the oxidation surface of the coating (Figure 3e2). The cross-sectional image (Figure 3e3) shows that the substrate has been severely oxidized. However, the interfacial bonding between the coating and the substrate remains relatively dense.

Figure 4 shows the GIXRD diffraction patterns of the coated sample after oxidation. For Al-0 coating, ZrO_2 and Nb_2O_5 were mainly detected, and the corresponding surface OM morphology showed that the coating had peeled off, causing the substrate to be directly exposed to the high-temperature water vapor environment and oxidized, thereby forming a large amount of ZrO_2. In contrast, for the Al-5, Al-10, Al-17, and Al-25 coatings, due to higher integrity, in addition to detecting simple oxides such as Al_2O_3, TiO_2, Cr_2O_3, ZrO_2, and Nb_2O_5 in the diffraction patterns, $CrTaO_4$ composite oxides were also observed. This phase has been reported to be formed by the further reaction of Cr_2O_3 and Ta_2O_5 [28]. With the increase in Al content, in the Al-17 and Al-25 coatings, $AlTaO_4$ was further detected, indicating that a high Al content may promote the formation of complex oxides. In addition, in the Al-10 coating with superior oxygen-blocking performance, intermetallic compounds Cr_5Al_8 and $ZrCr_2$ were identified, which may be related to the incomplete oxidation of the coating during the oxidation process.

Figure 5 shows the local magnified SEM morphology of the Al-0 coating after being held at 1000 °C in water vapor for 60 min. The results indicate that under the action of high-temperature water vapor, the coating was severely fragmented and completely separated from the substrate, and obvious cracks can also be observed in the independent remaining coating. The line scan in Figure 5b shows that the content of the O element gradually decreased from the outer layer to the inner layer of the coating, while other elements were relatively uniformly distributed within the coating area, and no significant enrichment was observed. The determination of the composition of different areas of the coating was carried out, and the result of site 1, shown in Table 3, indicates that the Cr content in the coating surface layer after oxidation was relatively high, indicating that Cr undergoes significant surface diffusion during oxidation. The composition analysis in site 2 shows that the content of the O element in this area was high, indicating that the coating has been completely oxidized. After removing the O element, the Cr content of the residual coating is lower than that of the deposited state coating due to surface diffusion, while the contents of Ti, Zr, Nb, and Ta change slightly, indicating that the diffusion behavior of these elements during oxidation is not obvious. In addition, the composition analysis result of site 3 shows that the substrate has been completely oxidized, and the coating fails to provide effective oxidation resistance protection. Overall, the RHEA coating without Al alloying undergoes severe fragmentation during high-temperature oxidation and cannot form a complete oxygen-blocking structure, resulting in the exposure of the substrate and intense oxidation.

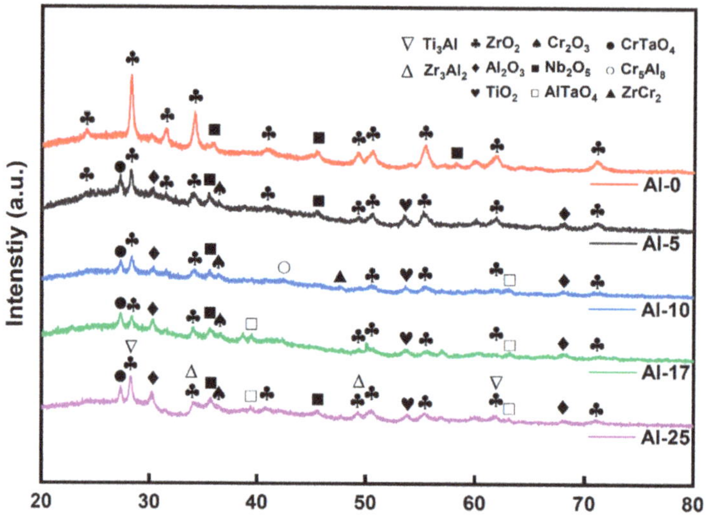

Figure 4. GIXRD spectra of the oxidized state AlTiCrZrNbTa coating.

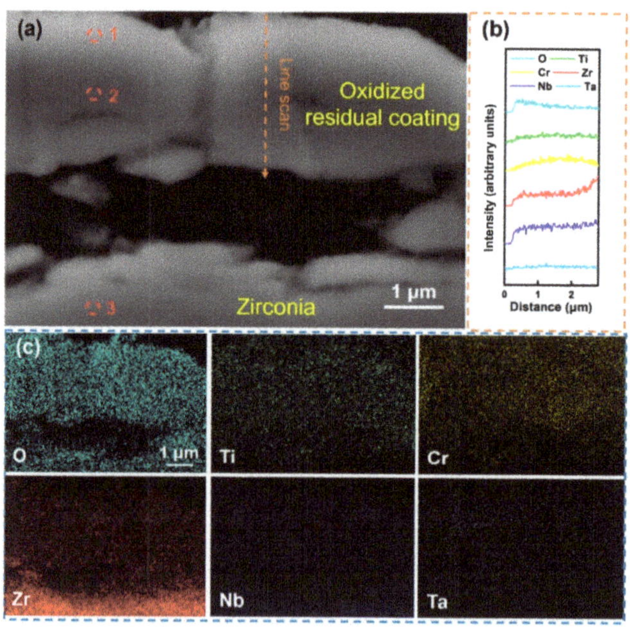

Figure 5. (**a**) Cross-sectional SEM images of the Al-0 coating after holding it in steam at 1000 °C for 60 min; (**b**) EDS line scan of (**a**) at the marked position; (**c**) EDS mapping of (**a**); the corresponding components in (**a**) are shown in Table 3.

Table 3. Components in the corresponding region in Figure 5.

Point	Element (at.%)					
	O	Ti	Cr	Zr	Nb	Ta
1	82.53	2.92	8.17	2.56	2.44	1.38
2	81.66	3.4	6.36	3.47	3.41	1.70
3	73.65	0.16	0.01	25.78	0.4	0

Figure 6a shows the local magnified morphology of the Al-5 coating after being kept in a water vapor environment at 1000 °C for 60 min. Compared with Al-0, the Al-5 coating maintained a complete structure and avoided severe cracking and peeling phenomena. The EDS analysis at site 1 in Table 4 confirms that the oxide layer was mainly composed of Cr_2O_3, accompanied by Al_2O_3 and TiO_2 inclusions. This indicates that Cr underwent significant external diffusion, which was conducive to the formation of the oxide layer. Meanwhile, the element distribution within the coating was relatively uniform, and site 2 shows that the O content in this area is lower than that of the surface oxide layer, indicating that the formation of the surface oxide layer improves the oxygen-blocking ability of the coating to a certain extent. However, although the Al-5 coating can maintain good integrity, there are still certain issues with its bonding performance to the substrate. In addition, site 3 indicates that the matrix has undergone severe oxidation. After long-term oxidation, the coating and the substrate became seriously separated.

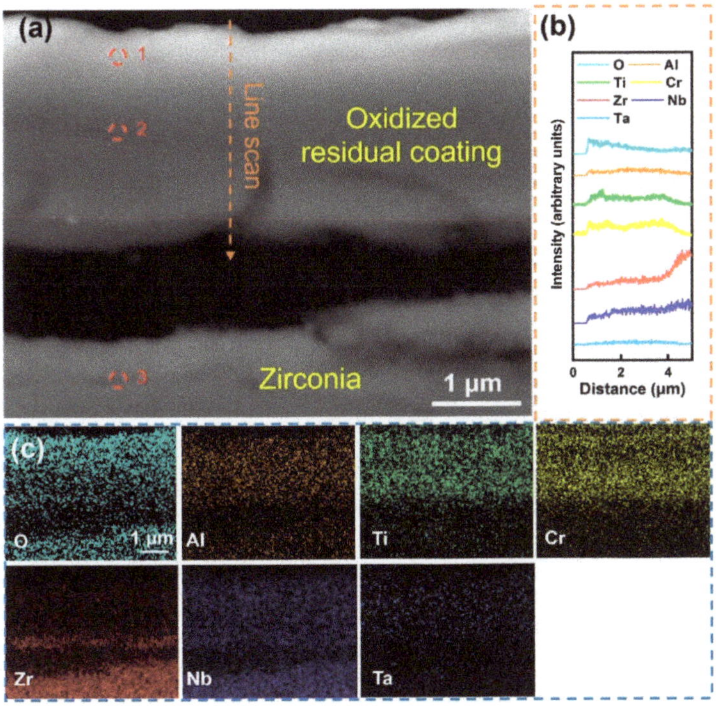

Figure 6. (**a**) Cross-sectional SEM images of the Al-5 coating after holding it in steam at 1000 °C for 60 min; (**b**) EDS line scan of the figure (**a**) at the marked position; (**c**) EDS mapping of (**a**); the corresponding components are shown in Table 4.

Table 4. Components in the corresponding region in Figure 6.

Point	Element (at.%)						
	O	Al	Ti	Cr	Zr	Nb	Ta
1	68.43	4.78	5.2	14.41	3.78	1.85	1.46
2	65.67	1.58	6.27	12.38	6.01	4.86	3.32
3	64.24	0	0.01	0.07	34.83	0.85	0

Figure 7 displays the cross-sectional morphology and elemental distribution of the Al-10 coating after oxidation in a 1000 °C water vapor environment for 60 min. A surface oxide layer with high oxygen content was observed (Figure 7c), enriched with Al, Ti, and Cr. Elemental line scanning (Figure 7b) also confirmed significant surface enrichment of Al, Ti, and Cr, suggesting their active participation in oxide layer formation during oxidation. Compositional analysis (Table 5, Site 1) revealed that the oxide layer primarily consists of Cr_2O_3, alloyed with Al and Ti. The oxide layer exhibited a dense, defect-free interface with the underlying coating, indicating strong adhesion and uniform elemental distribution within the residual coating. Notably, the residual coating (Site 2) showed markedly reduced oxygen content compared to low-Al counterparts, demonstrating enhanced oxygen-blocking capability. Furthermore, negligible oxygen diffusion into the substrate (Figure 7c) and a significant decrease in matrix oxygen content (Site 3) confirmed the coating's effective barrier against oxidative penetration.

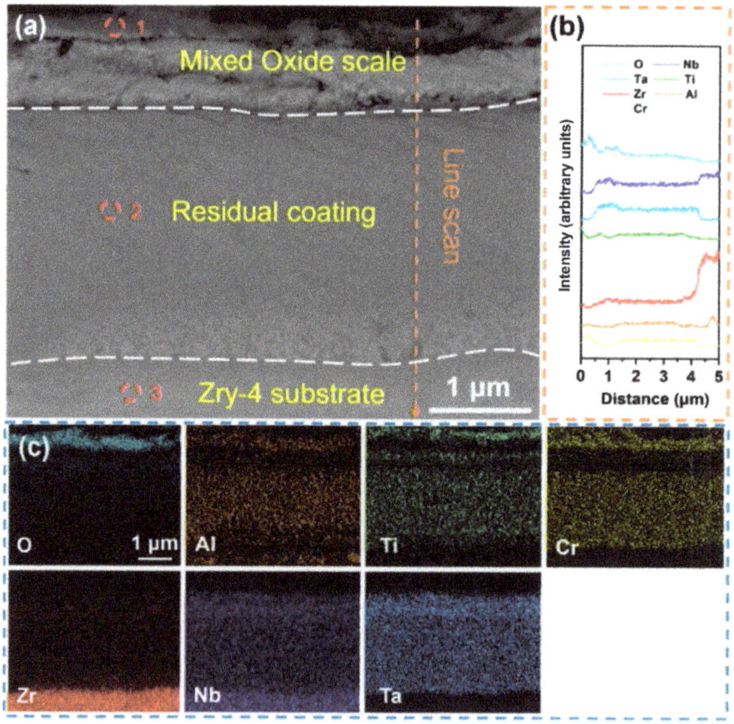

Figure 7. (**a**) Cross-sectional SEM images of the Al-10 coating after holding it in steam at 1000 °C for 60 min; (**b**) EDS line scan of (**a**) at the marked position; (**c**) EDS mapping of (**a**); the corresponding components in (**a**) are shown in Table 5.

Table 5. Components in the corresponding region in Figure 7.

Point	Element (at.%)						
	O	Al	Ti	Cr	Zr	Nb	Ta
1	59.85	6.14	10.26	19.28	2.76	1.55	0.16
2	31.42	6.76	9.57	27.95	9.14	10.79	4.37
3	17.35	1.56	0.45	0.81	75.29	4.54	0

The high-temperature water vapor oxidation behavior of the Al-10 coating indicated that the appropriate addition of Al can significantly enhance the oxidation resistance of the coating. However, as the Al content increased further, although the coating remained intact and exhibited strong bonding with the substrate, its oxygen-blocking performance was reduced. Figure 8 shows the cross-sectional morphology and element distribution of the Al-17 coating after being held in a water vapor environment at 1000 °C for 60 min. From the EDS mapping results in Figure 8c, it can be seen that the O content in the substrate has significantly increased, indicating that the coating failed to effectively prevent the diffusion of the O element into the substrate. The distribution of the O element in the coating was relatively obvious. The element analysis at site 1 in Table 6 shows that the coating surface mainly formed a mixed oxide composed of Al_2O_3, TiO_2, and Cr_2O_3. The element analysis at site 2 indicates that the coating has been totally oxidized. Although the Al-17 coating can maintain structural integrity, its oxygen-blocking ability is lower than that of the Al-10 coating. The excessive addition of Al may weaken the oxidation resistance of the coating, making O more likely to penetrate and oxidize the substrate (site 3).

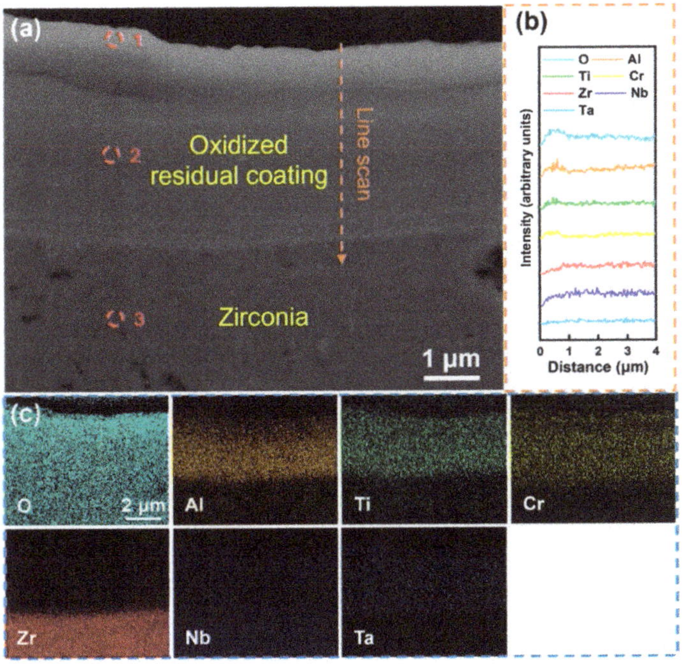

Figure 8. (**a**) Cross-sectional SEM images of the Al-17 coating after holding it in steam at 1000 °C for 60 min; (**b**) EDS line scan of (**a**) at the marked position; (**c**) EDS mapping of (**a**); the corresponding components in (**a**) are shown in Table 6.

Table 6. Components in the corresponding region in Figure 8.

Point	Element (at.%)						
	O	Al	Ti	Cr	Zr	Nb	Ta
1	68.63	6.54	6.06	13.59	2.23	1.96	1.17
2	66.68	5.65	4.17	10.32	5.42	5.76	2.45
3	64.48	1.69	0.58	0.32	29.13	0.8	0.31

Figure 9 shows the cross-sectional morphology and element distribution of the Al-25 coating after being maintained at 1000 °C in a high-temperature water vapor environment for 60 min. After oxidation, the coating still maintains good bonding with the substrate. However, as can be seen from the distribution of the O element in Figure 9c, the substrate has been completely oxidized, and no obvious surface oxide layer has been formed. The line scan results in Figure 9b indicate that there are Al element-enriched regions on the coating surface. The component analysis in Table 7 at site 1 shows that the contents of Al, Ti, and Cr on the coating surface have slightly increased, which is mainly attributed to the surface diffusion effect of elements in the early stage of oxidation. The component analysis at site 2 indicates that the content of the Al element in the coating interior is relatively high, and compared with the deposited-state coating, it has not undergone large-scale diffusion to the surface. However, the component analysis at site 3 shows that the substrate has been completely oxidized, and the coating failed to effectively prevent the diffusion of the O element, ultimately leading to the failure of the coating. Overall, although the Al-25 coating still maintains structural integrity during the high-temperature oxidation process, its oxygen-blocking ability has not been effectively enhanced, and the substrate has suffered severe oxidation, indicating that excessive Al alloying may have an adverse effect on the oxidation resistance of the coating.

Table 7. Components in the corresponding region in Figure 9.

Point	Element (at.%)						
	O	Al	Ti	Cr	Zr	Nb	Ta
1	82.8	6.18	4.66	3.95	0.6	1.1	0.8
2	77.6	6.5	4.4	2.9	3.6	3.7	1.3
3	76.28	0	0.04	0	23.49	0	0.19

Through the high-temperature oxidation experiments of RHEA coatings alloyed with different Al contents, it is known that the addition of active element Al can effectively enhance the oxidation resistance of the coating. However, in order to ensure that the coating can form a stable oxygen-blocking structure during the oxidation process and maintain a good bonding with the substrate to achieve long-term protection, the Al content should be controlled within a reasonable range of 9–14 at.%. To further verify the oxygen-blocking effect of Al on RHEA coatings under high-temperature conditions, a water vapor oxidation experiment at a higher temperature was conducted on the Al-10 coating with better performance. Figure 10a shows the cross-sectional morphology of the Al-10 coating after being held at 1100 °C for 30 min. According to the distribution of the O element in Figure 10b, an oxide layer was formed on the coating surface, which was mainly composed of Al_2O_3, TiO_2, and Cr_2O_3. The experimental results show that the Al-10 coating can still effectively prevent oxygen diffusion under higher temperature environments, demonstrating excellent oxidation resistance, further verifying the positive role of appropriate Al alloying on the oxidation resistance of RHEA coatings.

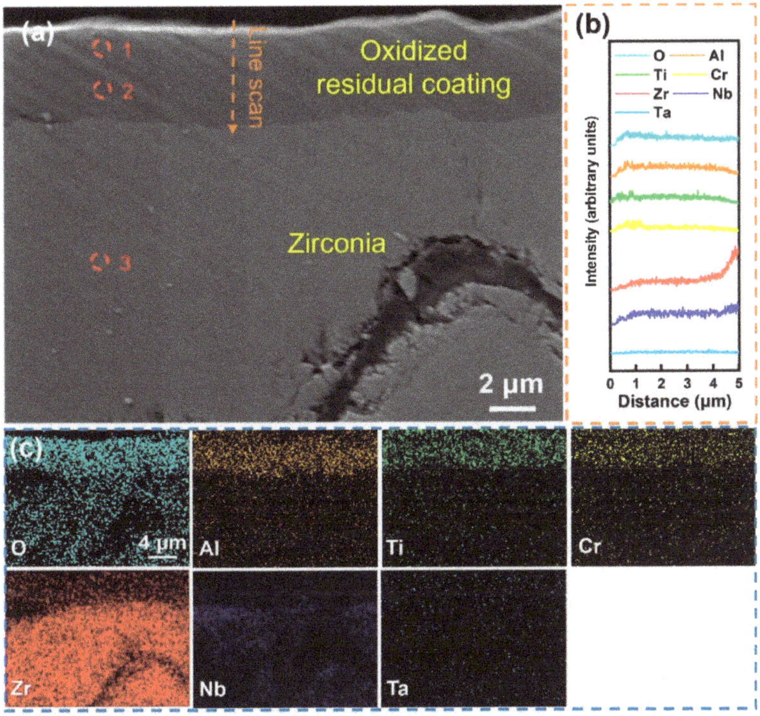

Figure 9. (**a**) Cross-sectional SEM images of the Al-25 coating after holding it in steam at 1000 °C for 60 min; (**b**) EDS line scan of (**a**) at the marked position; (**c**) EDS mapping of (**a**); the corresponding components in (**a**) are shown in Table 7.

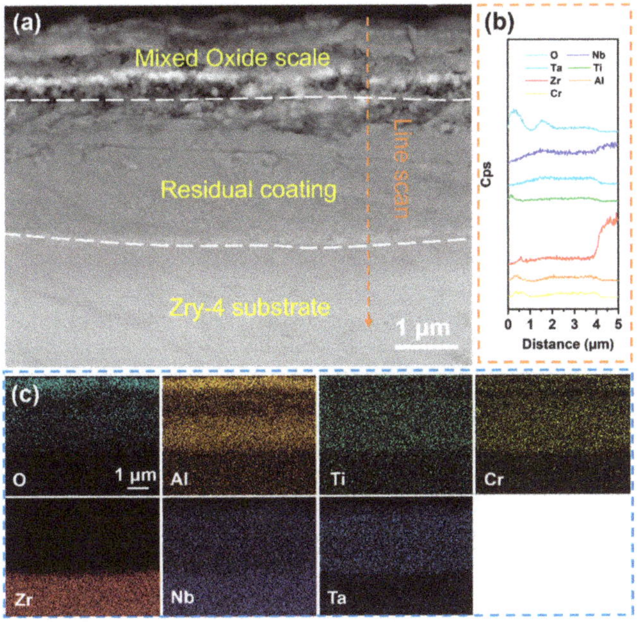

Figure 10. (**a**) Cross-sectional SEM images of the Al-10 coating after holding it in steam at 1100 °C for 30 min; (**b**) EDS line scan of (**a**) at the marked position; (**c**) EDS mapping of (**a**).

The refractory high-entropy alloy coating without Al alloying cracked and directly peeled off from the substrate surface, indicating poor bonding with the substrate. At the same time, due to the low plasticity of refractory metal oxides, brittle fracture is prone to occur under external impact, further exacerbating the coating detachment. Al alloying significantly improves coating-substrate bonding, as demonstrated by indentation tests following 1000 °C steam exposure (Figure 11). Unalloyed coatings exhibited catastrophic delamination and brittle fracture under mechanical stress, attributable to the inherent brittleness of refractory metal oxides. In contrast, Al-modified coatings (e.g., Al-5 and Al-10) maintained structural integrity despite localized microcracking. The Al-10 coating, in particular, displayed superior adhesion with minimal internal cracking, a result of its reduced oxygen content and stabilized oxide layer. These findings align with prior studies [38], wherein Al alloying in high-entropy alloys (HEAs) enhanced interfacial cohesion through improved hardness, fracture toughness, and bending resistance. However, excessive Al addition ($\geq$17 at.%) triggered outward crack propagation, albeit without complete structural failure, highlighting Al's dual role in optimizing adhesion and oxidation resistance. This underscores the necessity of precise Al content regulation to achieve balanced mechanical and anti-oxidative performance in extreme environments.

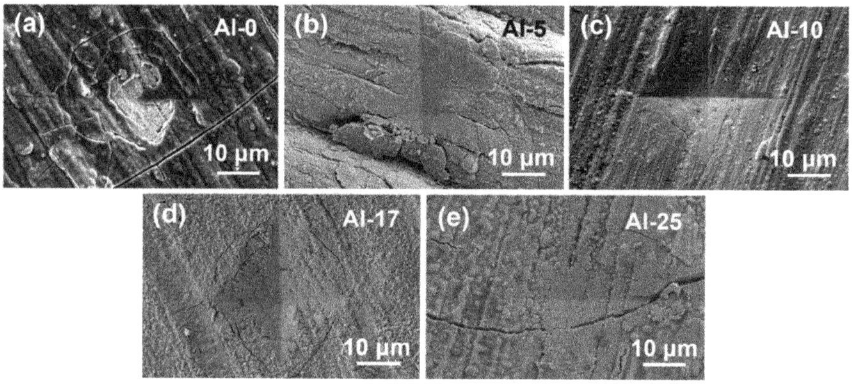

Figure 11. SEM image of the surface indentation of the oxidized state coatings: (**a**) Al-0; (**b**) Al-5; (**c**) Al-10; (**d**) Al-17; (**e**) Al-25.

The absence of Al alloying in refractory high-entropy alloy coatings results in fragmented oxidation and substrate exposure due to the inability to form a protective oxide barrier. Al addition promotes surface oxide stability, with the Al-10 coating achieving optimal performance through a dense, continuous Al_2O_3-TiO_2-Cr_2O_3 composite layer during high-temperature steam oxidation. Thermodynamic analysis (Figure 12a) reveals that Al_2O_3 and TiO_2 preferentially nucleate due to their lower Gibbs free energy (ΔG) [39–46], while Cr enrichment compensates for its marginally higher ΔG to stabilize Cr_2O_3 [47]. Despite ZrO_2's thermodynamic stability, its sluggish diffusion kinetics and large atomic size impede surface layer formation. Excessive Al ($\geq$17 at.%) induces Al-Ti/Al-Zr intermetallic phases (Figure 12b), which restrict Al diffusion and disrupt oxide continuity, as explained by the negative mixing enthalpy favoring stable Al-Zr/Ti bonding [48,49]. Concurrently, insufficient Ti/Cr surface enrichment and Al-rich phase accumulation degrade oxidation resistance. These mechanisms underscore the necessity of precise Al content control to balance oxide nucleation kinetics and elemental mobility (Figure 12c), ensuring robust oxygen-blocking performance in extreme environments.

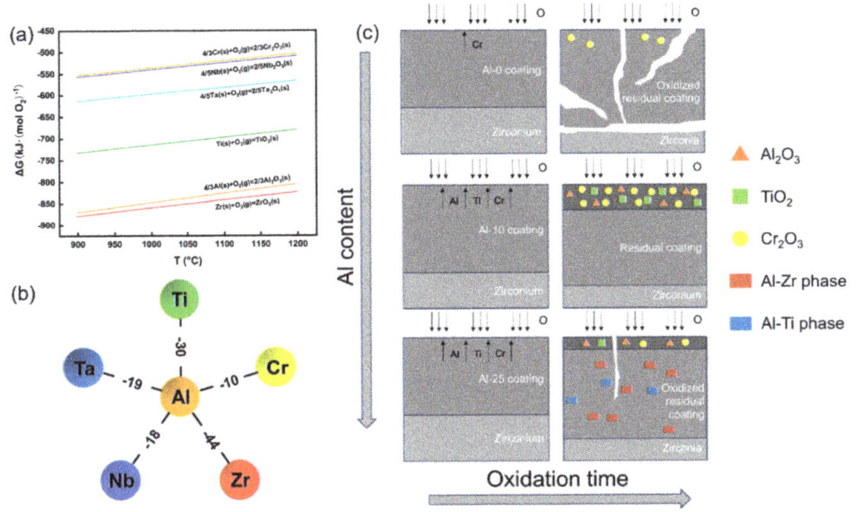

Figure 12. (**a**) Formation free energy of stable oxides of elements in AlTiCrZrNbTa coating at 900–1200 °C; (**b**) heat of mixing of Al with other components in the coating; (**c**) schematic diagram of the oxidation process of AlTiCrZrNbTa coatings with different Al contents during isothermal oxidation in water vapor environment at 1000 °C.

Compared our results with previous studies [50,51]. The literature reports that increasing Al content in alloys enhances the formation of a thicker and more continuous Al_2O_3 layer, which protects the substrate from atmospheric oxidation. Our findings differ, as we focus on coatings rather than bulk alloys. In coatings, the compatibility between the residual layer and the oxide layer must be considered. With increasing Al content, thermal expansion mismatch and the accumulation of intermetallics in the residual coating lead to oxide layer cracking and discontinuity. Furthermore, the high oxygen solubility of these intermetallics reduces the overall oxidation resistance of the coating.

4. Conclusions

This study systematically evaluates the oxidation resistance of AlTiCrZrNbTa refractory high-entropy alloy (RHEA) coatings on Zry-4 substrates under high-temperature steam environments, with a focus on the critical role of Al alloying. Key findings reveal that Al content profoundly influences coating performance through distinct mechanisms. At an optimal Al concentration of 10 at.%, the coating forms a dense, continuous composite oxide layer (Al_2O_3-TiO_2-Cr_2O_3) during oxidation, which effectively suppresses oxygen diffusion into the substrate and maintains robust interfacial adhesion. Thermodynamic analysis confirms the preferential nucleation of Al_2O_3 and TiO_2 due to their low Gibbs free energy, while Cr enrichment enhances oxide layer stability. Mechanical indentation tests further validate the structural integrity of the Al-10 coating, with minimal post-oxidation cracking. Conversely, excessive Al alloying (≥17 at.%) promotes the formation of detrimental Al-Ti and Al-Zr intermetallic phases, which restrict reactive element diffusion and disrupt the continuity of the protective oxide layer, leading to accelerated substrate oxidation. These findings establish a critical Al threshold for balancing oxide layer formation and elemental mobility. The Al-10 coating demonstrates exceptional stability under extreme conditions, highlighting its potential as a durable protective layer for zirconium-based nuclear fuel claddings. This work provides a compositional design framework for optimizing oxidation resistance in RHEA coatings, advancing the development of accident-tolerant fuel systems for enhanced nuclear safety.

Author Contributions: M.G.: conceived the research question, conducted the investigation, wrote the original draft, and was responsible for formal analysis. C.C., B.S. and J.G.: handled the methodological design, project administration, funding acquisition, and participated in the writing—review and editing. J.H.: methodology, visualization, and funding acquisition. G.C.: supervised the entire project, contributed to the methodological design, project administration, and funding acquisition, and participated in the writing—review and editing. All authors have read and agreed to the published version of the manuscript.

Funding: This research was funded by Frontier Exploration Project of Longmen Laboratory (LMOYT-SKT009), Natural Science Foundation of Henan [232300421013], National Natural Science Foundation of China [52403406, 52171082], Education Department of Henan Province [23A430014], Science and Technology Department of Henan Province [232102230043, 232301420001, 241111221500].

Institutional Review Board Statement: Not applicable.

Informed Consent Statement: Not applicable.

Data Availability Statement: The original contributions presented in this study are included in the article Further inquiries can be directed to the corresponding author.

Conflicts of Interest: The authors declare that they have no known competing financial interests or personal relationships that could have appeared to influence the work reported in this paper.

References

1. Binns, C.; Low, W.Y. Fukushima Disaster: Public Health Issues. *Asia Pac. J. Public Health* **2017**, *29*, 5S–6S. [CrossRef]
2. Gu, Z. History review of nuclear reactor safety. *Ann. Nucl. Energy* **2018**, *120*, 682–690. [CrossRef]
3. Rogner, H.-H. World outlook for nuclear power. *Energy Strategy Rev.* **2013**, *1*, 291–295. [CrossRef]
4. Karpyuk, L.A.; Savchenko, A.M.; Leont'eva-Smirnova, M.V.; Kulakov, G.V.; Konovalov, Y.V. Steel Cladding for VVER Fuel Pins in the Context of Accident-Tolerant Fuel: Prospects. *At. Energy* **2020**, *128*, 218–222. [CrossRef]
5. Jin, D.; Guo, Y.; Gao, Z.; Zhou, Y.; Wang, X.; Guo, F.; Zhao, X.; Xiao, P. Investigation on the oxidation and corrosion behaviors of FeCrZr alloy as a protective material for Zr cladding. *J. Alloys Compd.* **2018**, *753*, 532–542. [CrossRef]
6. Was, G.S.; Petti, D.; Ukai, S.; Zinkle, S. Materials for future nuclear energy systems. *J. Nucl. Mater.* **2019**, *527*, 151837. [CrossRef]
7. Yang, L.; Zhou, Z.; Yang, R.; Wang, J.; Chen, M.; Qiao, Y.; Zhu, S.; Wang, F. Effect of Al and Cr on the oxidation behavior of nanocrystalline coatings at 1050 °C. *Corros. Sci.* **2022**, *200*, 110191. [CrossRef]
8. Li, G.; Liu, Y.; Zhang, Y.; Wang, J.; Chen, M.; Qiao, Y.; Zhu, S.; Wang, F. High Temperature Anti-Oxidation Behavior and Mechanical Property of Radio Frequency Magnetron Sputtered Cr Coating. *Metals* **2020**, *10*, 1509. [CrossRef]
9. Hong, J.; Ahn, D.; Jang, D.; Oh, H.; Kim, J.; Kim, H.; Ševeček, M. Measurement of local mechanical properties for Cr-coated accident tolerant fuel cladding. *J. Nucl. Mater.* **2023**, *579*, 154407. [CrossRef]
10. Li, Y.; Lv, H.; Li, Y.; Fan, N. Effect of Pre-Oxidation on High-Temperature Oxidation Behavior of Al-Si Coating on Nickel-Based Superalloy. *Materials* **2022**, *15*, 7440. [CrossRef]
11. Yang, J.; Zhou, M.; Lv, L.; Zhou, Y.; Li, Q.; Liu, N.; Yang, J. Influence of Si addition on the microstructure, mechanical and lead-bismuth eutectic corrosion properties of an amorphous AlCrFeMoTiSix high-entropy alloy coating. *Intermetallics* **2022**, *148*, 107649. [CrossRef]
12. Al-Olayyan, Y.; Fuchs, G.; Baney, R.; Tulenko, J. The effect of Zircaloy-4 substrate surface condition on the adhesion strength and corrosion of SiC coatings. *J. Nucl. Mater.* **2005**, *346*, 109–119. [CrossRef]
13. Yeom, H.; Hauch, B.; Cao, G.; Garcia-Diaz, B.; Martinez-Rodriguez, M.; Colon-Mercado, H.; Olson, L.; Sridharan, K. Laser surface annealing and characterization of Ti_2AlC plasma vapor deposition coating on zirconium-alloy substrate. *Thin Solid Film.* **2016**, *615*, 202–209. [CrossRef]
14. Tang, C.; Stueber, M.; Seifert, H.J.; Steinbrueck, M. Protective coatings on zirconium-based alloys as accident-tolerant fuel (ATF) claddings. *Corros. Rev.* **2017**, *35*, 141–165. [CrossRef]
15. Huang, Y.; Lin, J.; Zhang, H. Effect of Si_3N_4 content on microstructures and antioxidant properties of $MoSi_2/Si_3N_4$ composite coatings on Mo substrate. *Ceram. Int.* **2015**, *41*, 13903–13907. [CrossRef]
16. Yeh, J.W.; Chen, S.K.; Lin, S.J.; Gan, J.Y.; Chin, T.S.; Shun, T.T.; Tsau, C.H.; Chang, S.Y. Nanostructured High-Entropy Alloys with Multiple Principal Elements: Novel Alloy Design Concepts and Outcomes. *Adv. Eng. Mater.* **2004**, *6*, 299–303. [CrossRef]

17. Tao, Z.; Wang, P.; Wang, C.; Ma, Z.; Zhang, Y.; Xue, F.; Bai, G.; Yuan, Y.; Lan, R. Design and characterisation of $AlCrFeCuNb_x$ alloys for accident-tolerant fuel cladding. *J. Alloys Compd.* **2021**, *859*, 157805. [CrossRef]
18. Butler, T.M.; Chaput, K.J.; Dietrich, J.R.; Senkov, O.N. High temperature oxidation behaviors of equimolar NbTiZrV and NbTiZrCr refractory complex concentrated alloys (RCCAs). *J. Alloys Compd.* **2017**, *729*, 1004–1019. [CrossRef]
19. Kumar, S. Comprehensive review on high entropy alloy-based coating. *Surf. Coat. Technol.* **2024**, *477*, 130327. [CrossRef]
20. Senkov, O.N.; Wilks, G.B.; Miracle, D.B.; Chuang, C.P.; Liaw, P.K. Refractory high-entropy alloys. *Intermetallics* **2010**, *18*, 1758–1765. [CrossRef]
21. Song, H.; Tian, F.; Wang, D. Thermodynamic properties of refractory high entropy alloys. *J. Alloys Compd.* **2016**, *682*, 773–777. [CrossRef]
22. Jiang, W.; Wang, X.; Kang, H.; Jiang, B.; Dong, D.; Wang, Y.; Zhu, D. Microstructure and mechanical properties of AlNbTiVZr system refractory high entropy alloys. *J. Alloys Compd.* **2022**, *925*, 166767. [CrossRef]
23. Ding, X.; Wang, W.; Zhang, H.; Tian, X.; Luo, L.; Wu, Y.; Yao, J. Study on Microstructure and High Temperature Stability of $WTaVTiZr_x$ Refractory High Entropy Alloy Prepared by Laser Cladding. *Entropy* **2024**, *26*, 73. [CrossRef]
24. Zong, Y.; Hashimoto, N.; Oka, H. Study on irradiation effects of refractory bcc high-entropy alloy. *Nucl. Mater. Energy* **2022**, *31*, 101158. [CrossRef]
25. Wang, B.; Yang, C.; Shu, D.; Sun, B. A Review of Irradiation-Tolerant Refractory High-Entropy Alloys. *Metals* **2023**, *14*, 45. [CrossRef]
26. Sadeghilaridjani, M.; Ayyagari, A.; Muskeri, S.; Hasannaeimi, V.; Salloom, R.; Chen, W.-Y.; Mukherjee, S. Ion irradiation response and mechanical behavior of reduced activity high entropy alloy. *J. Nucl. Mater.* **2020**, *529*, 151955. [CrossRef]
27. Pu, G.; Sun, S.; Wang, S.; Gan, L.; Chen, S.; Ye, Z.; Huang, Z.; Zheng, J.; Wang, Z.; Yang, C.; et al. A comparative study of irradiation response in amorphous TaTiWVCr refractory high entropy alloy with the counterpart of tungsten films. *Intermetallics* **2023**, *156*, 107850. [CrossRef]
28. Müller, F.; Gorr, B.; Christ, H.-J.; Müller, J.; Butz, B.; Chen, H.; Kauffmann, A.; Heilmaier, M. On the oxidation mechanism of refractory high entropy alloys. *Corros. Sci.* **2019**, *159*, 108161. [CrossRef]
29. Anber, E.A.; Beaudry, D.; Brandenburg, C.; Lech, S.; Backman, L.; Foley, D.L.; Wang, E.L.; Waters, M.J.; Perriere, L.; Couzinie, J.P.; et al. Oxidation resistance of Al-containing refractory high-entropy alloys. *Scr. Mater.* **2024**, *244*, 115997. [CrossRef]
30. Pei, X.; Du, Y.; Wang, H.; Li, T.; Hu, M.; Wang, H.; Liu, W. Effects of Al/Si on the oxidation behavior of a TiZrV0.5Nb0.5 refractory high entropy alloy at 1000 °C. *Corros. Sci.* **2023**, *224*, 111527. [CrossRef]
31. Gorr, B.; Schellert, S.; Müller, F.; Christ, H.J.; Kauffmann, A.; Heilmaier, M. Current Status of Research on the Oxidation Behavior of Refractory High Entropy Alloys. *Adv. Eng. Mater.* **2021**, *23*, 2001047. [CrossRef]
32. Zhang, Y.; Zhou, Y.J.; Lin, J.P.; Chen, G.L.; Liaw, P.K. Solid-solution phase formation rules for multi-component alloys. *Adv. Eng. Mater.* **2008**, *10*, 534–538. [CrossRef]
33. Zhao, S.; Liu, C.; Yang, J.; Zhang, W.; He, L.; Zhang, R.; Yang, H.; Wang, J.; Long, J.; Chang, H. Mechanical and high-temperature corrosion properties of AlTiCrNiTa high entropy alloy coating prepared by magnetron sputtering for accident-tolerant fuel cladding. *Surf. Coat. Technol.* **2021**, *417*, 127228. [CrossRef]
34. He, L.; Liu, C.; Zhao, S.; Shu, C.; Yang, J.; Liu, H.; Zhang, W.; Lin, J.; Long, J.; Chang, H. Microstructure, mechanical and corrosion properties of high hardness TiVZrCrAl HEA coatings prepared by magnetron sputtering. *Surf. Coat. Technol.* **2022**, *441*, 128532. [CrossRef]
35. Li, Z.; Li, Y.; Hong, Y.; Wu, J.; Qin, W.; Wu, X. Effect of Si on the structure and oxidation behavior of Cr-Si coatings on Zry-4 under high temperature steam environment. *J. Nucl. Mater.* **2024**, *597*, 155126. [CrossRef]
36. Kashkarov, E.B.; Sidelev, D.V.; Pushilina, N.S.; Yang, J.; Tang, C.; Steinbrueck, M. Influence of coating parameters on oxidation behavior of Cr-coated zirconium alloy for accident tolerant fuel claddings. *Corros. Sci.* **2022**, *203*, 110359. [CrossRef]
37. Zhu, Y.; Tao, Z.; Gu, X.; Wang, C.; Lan, R. Enhanced accident tolerance properties of $AlCrCuFeMoNb_x$ high entropy coating for nuclear fuel cladding. *J. Alloys Compd.* **2025**, *1010*, 177206. [CrossRef]
38. Kim, D.H.; Hong, J.; Kim, H.D.; Kim, J.; Guim, H. Measurement and investigation of mechanical properties of a partial oxide dispersion strengthened Zircaloy-4 tube via an instrumented indentation test. *Mater. Today Commun.* **2021**, *27*, 102210. [CrossRef]
39. Rossnagel, S.M. Magnetron sputtering. *J. Vac. Sci. Technol. A* **2020**, *38*, 060805. [CrossRef]
40. Bai, H.; Li, J.; Gao, J.; Ni, J.; Bai, Y.; Jian, J.; Zhao, L.; Bai, B.; Cai, Z.; He, J.; et al. Comparison of CrN Coatings Prepared Using High-Power Impulse Magnetron Sputtering and Direct current magnetron sputtering. *Materials* **2023**, *16*, 6303.
41. Gu, B.; Wen, G.; Zhang, B.; Petr, J.; Wagner, T. 2D and thin-film copper synthesized via magnetron sputtering. *MRS Commun.* **2024**, *14*, 317–322. [CrossRef]
42. Sasani Ghamsari, M. Development of Thin Film Fabrication Using Magnetron Sputtering. *Metals* **2023**, *13*, 963. [CrossRef]
43. Padamata, S.K.; Yasinskiy, A.; Yanov, V.; Saevarsdottir, G. Magnetron Sputtering High-Entropy Alloy Coatings: A Mini-Review. *Metals* **2022**, *12*, 319. [CrossRef]

44. Alat, E.; Motta, A.T.; Comstock, R.J.; Partezana, J.M.; Wolfe, D.E. Ceramic coating for corrosion (c3) resistance of nuclear fuel cladding. *Surf. Coat. Technol.* **2015**, *281*, 133–143. [CrossRef]
45. Alat, E.; Motta, A.T.; Comstock, R.J.; Partezana, J.M.; Wolfe, D.E. Multilayer (TiN, TiAlN) ceramic coatings for nuclear fuel cladding. *J. Nucl. Mater.* **2016**, *478*, 236–244. [CrossRef]
46. Jayaraj, J.; Thirathipviwat, P.; Han, J.; Gebert, A. Microstructure, mechanical and thermal oxidation behavior of AlNbTiZr high entropy alloy. *Intermetallics* **2018**, *100*, 9–19. [CrossRef]
47. Zeng, S.; Li, J.F.; Chen, C.; Meng, Y.; Zhu, C.W.; Han, X.C.; Bao, Y.W.; Zhang, H.B. Improved oxidation resistance of Cr-Si coated Zircaloy with an in-situ formed Zr_2Si diffusion barrier. *NPJ Mater. Degrad.* **2023**, *7*, 56. [CrossRef]
48. Takeuchi, A.; Inoue, A. Classification of bulk metallic glasses by atomic size difference, heat of mixing and period of constituent elements and its application to characterization of the main alloying element. *Mater. Trans.* **2005**, *46*, 2817–2829. [CrossRef]
49. Gu, X.; Luan, H.; Yang, X.; Wang, X.; Fang, K.; Li, J.; Jia, Y.; Yao, K.; Zhang, Z.; Chen, N. Formation and Properties of Amorphous Multi-Component $(CrFeMoNbZr)O_x$ Thin Films. *Metals* **2020**, *10*, 599. [CrossRef]
50. Wei, Y.; Fu, Y.; Pan, Z.; Ma, Y.; Cheng, H.; Zhao, Q.; Luo, H.; Li, X. Influencing factors and mechanism of high-temperature oxidation of high-entropy alloys: A review. *Int. J. Miner. Metall. Mater.* **2021**, *28*, 915–930. [CrossRef]
51. Xin, H.; Yang, J.; Mao, J.; Chen, Q.; Yang, J.; Zhang, W.; Ning, Z.; Teng, C.; Ma, C.; Wu, L.; et al. Effect of Al concentration on microstructure and properties of AlNbTiZr medium-entropy alloy coatings. *Materials* **2021**, *14*, 7661. [CrossRef] [PubMed]

materials

MDPI

Article

Effect of Si Addition on Structure and Corrosion Resistance of FeCoNiCr High-Entropy Alloy Coating

Wenqiang Li [1,2], Jie Lian [3], Dengfeng Wang [3], Suo Zhang [3,*], Chengfu Han [1], Zhenyu Du [1] and Fushan Li [1,*]

[1] School of Materials Science and Engineering, Zhengzhou University, Zhengzhou 450001, China
[2] Henan Coal Science Academy Wear-Resistant Technology Co., Ltd., Zhengzhou 450001, China
[3] School of Mechanical and Electrical Engineering & Intelligent Manufacturing, Henan Open University, Zhengzhou 450008, China
* Correspondence: author: zhangsuo@haou.edu.cn (S.Z.); fsli@zzu.edu.cn (F.L.)

Abstract: In this study, $Fe_{60}Co_{10-x}Ni_{15}Cr_{15}Si_x$ (x = 0, 4, and 8) powders were successfully prepared using the aerosol method and employed to produce high-entropy coatings on Q235 steel via laser cladding. The microstructure and phase composition of the coatings were analyzed using scanning electron microscopy, energy-dispersive X-ray spectroscopy, and X-ray diffraction. Corrosion resistance and potential were evaluated through electrochemical analysis and Kelvin probe force microscopy. The results show that the $Fe_{60}Co_{10-x}Ni_{15}Cr_{15}Si_x$ coatings exhibit excellent metallurgical bonding with no visible porosity or cracks. The coating primarily consists of an FCC structure; however, as the Si content increases, the structure transitions to a mixed FCC + BCC phase. The addition of Si also refines the grain size in the alloy system. Electrochemical analysis reveals that the Si0 and Si4 coatings exhibit similar corrosion behavior, while the Si8 coating shows a significant drop in corrosion potential, reducing its corrosion resistance. As the Si content increases, grain refinement leads to more grain boundaries, but the corrosion resistance decreases due to the lower corrosion performance of Si compared to Co. Considering both cost and corrosion resistance, the Si4 coating offers a balance of low cost and excellent corrosion resistance.

Keywords: laser cladding; high-entropy alloy; microstructure analysis; corrosion resistance; coating

1. Introduction

High-entropy alloys (HEAs), which leverage the principle of configurational entropy, challenge the traditional alloy design paradigm reliant on a single principal element. HEAs offer superior properties such as high hardness, excellent wear and corrosion resistance, superior thermal stability, and superior irradiation resistance compared to conventional alloys [1,2]. However, HEAs have some drawbacks, including higher costs due to the use of expensive elements such as niobium (Nb) [3], vanadium (V) [4], tungsten (W), nickel (Ni), titanium (Ti) [5], and cobalt (Co) [6]. The widespread application of HEAs is further hindered by the high costs associated with their casting-based production methods.

The durability and performance of materials in various environments largely depend on surface quality, functionality, and mechanical properties [7]. To improve surface properties, numerous surface modification techniques are employed, including laser cladding, thermal spraying, magnetron sputtering, and plasma transfer arc cladding. Among these, laser cladding (LC) has been extensively studied and widely applied due to its ability to achieve controlled heat-affected zones (HAZs) with minimal dilution [8–14]. The high cooling rates (10^4–10^6 °C/s) in laser cladding promote structure refinement and prevent component segregation, making HEA-based coatings increasingly popular. HEA coatings are widely used in engineering and industrial applications for erosion resistance, corrosion protection, and stress mitigation [15]. Consequently, LC has emerged as a promising method for fabricating high-quality coatings with exceptional properties while addressing factors like processing complexity and environmental protection.

Currently, numerous laser-cladded HEA systems have been developed and extensively investigated. Among these, FeCoNiCr medium-entropy alloys (MEAs) with a face-centered cubic (FCC) structure are widely used due to their excellent solid solution capability [16–18]. Although FeCoCrNi MEAs exhibit good ductility, their hardness and strength are often insufficient for specific surface modification applications.

Recent studies have demonstrated that adding elements such as Mo, Al, Ti, Mn, and W to FeCoNiCr MEAs can improve their corrosion and wear resistance [19–23]. In addition, elements such as silicon (Si), chromium (Cr), nickel (Ni), and molybdenum (Mo) are frequently added to regulate phase structure refinement and enhance corrosion resistance in HEAs. These elements often contribute to passivation film formation, improving corrosion resistance. For example, the addition of Si element influences the thermodynamic state and structure of FeCoNiCr systems, as the approximate structure of this HEA system can be predicted based on its basic alloy parameters, including solution interaction (Ω), atomic size difference (δ), and valence electron concentration (VEC) [24,25]. The valence electron concentration of an equimolar FeCoCrNi system was calculated to be 8.25 (FCC structure) [26]. Yang et al. [27] reported that a solid solution is favored if $\delta < 6\%$ and $\Omega > 1$. Si element in the FCC phase of high-entropy alloys can enhance solid solution strengthening and reduce the size of the grain, resulting in an increase in the strength and toughness of the alloy [28,29]. The presence of Si element in the HEAs has been observed to enhance oxidation resistance by facilitating the formation of protective oxide layers on the surface, rendering them highly suitable for high-temperature applications [30].

Current research focuses on the phase transformation mechanisms of FeCoNiCr MEAs during heat treatment and deformation, with limited exploration of their low-cost application as coatings for corrosion resistance. Si element had a significant effect on the increase in the microhardness and wear resistance in FeCoCrNiSi HEAs, which is beneficial for its application in coating [31]. $Fe_{60}Co_{15}Ni_{15}Cr_{10}$ and $Fe_{60}Ni_{15}Cr_{15}Co_{10}$ medium-entropy alloys are composed primarily of a near-single FCC phase, which exhibits excellent plasticity, relatively low strength or hardness [16,32]. In order to reduce the raw material cost of these Fe-rich medium-entropy alloys and enhance their hardness, Si elements are added to this alloy system. In this study, $Fe_{60}Co_{10-x}Ni_{15}Cr_{15}Si_x$ (x = 0, 4, and 8) high-entropy alloys were prepared based on these FeCoNiCr systems. Laser cladding technology was used to create coatings on Q235 steel, aiming to investigate the structural and corrosion resistance mechanisms. This study seeks to develop a cost-effective HEA coating with high corrosion resistance, with the goal of contributing to the successful application of HEAs.

2. Experimental Section

2.1. Sample Preparation

In this study, the Q235 alloy (provided by Zhongnuo New Material (Beijing) Technology Co., Ltd., Beijing, China, composition shown in Table 1) was used as the base material, and Fe, Co, Ni, Cr, and Si with 99.9% purity (provided by Zhongnuo New Material (Beijing) Technology Co., Ltd., purity, shape, and size shown in Table 2) were selected to prepare the alloy composition $Fe_{60}Co_{10-x}Ni_{15}Cr_{15}Si_x$ (x = 0, 4, and 8). The alloy was melted in a vacuum arc furnace (Model: WK II, Physcience Opto-electronics Co., Ltd., Beijing, China), and 10 mm × 10 mm × 100 mm square columns were formed via suction casting. The specific procedure involves placing the raw materials, which have been formulated according to the nominal composition specified in the design, into the crucible of the copper mold within the vacuum arc furnace, followed by evacuation of the chamber. Once the vacuum level within the chamber reaches below 5×10^{-3} Pa, high-purity (99.9%) argon gas is introduced as a shielding medium. Subsequently, the high-purity titanium ingot is melted using the vacuum arc furnace to eliminate any residual oxygen present in the chamber. To ensure homogeneity of the alloy composition during the melting process, the alloy is remelted a minimum of five times.

Table 1. Composition of the Q235 alloy.

Elements	C	Si	Mn	S	P	Fe
Content (wt.%)	0.150	0.200	0.460	0.022	0.012	Balance

Table 2. The purity, shape, and size of high-purity Fe, Co, Ni, Cr, and Si materials for this study.

Element	Purity (wt.%)	Shape	Size
Fe	99.9%	Block	3 × 3 mm
Co	99.95%	Block	5–10 mm
Ni	99.9%	Block	3 × 3 mm
Cr	99.95%	Block	1–5 mm
Si	99.999%	Block	1–6 mm

Subsequently, $Fe_{60}Co_{10-x}Ni_{15}Cr_{15}Si_x$ HEA powder was prepared using aero-chemical equipment (Model: XGE-500, made by Shenyang Seagull Technology Co., Ltd., Shenyang, China) for the cladding material. The alloy master ingot (~80 g) is placed into a quartz tube with a 1 mm diameter at its lower end. The tube is subsequently evacuated to a pressure of 10^{-3} Pa, followed by the introduction of argon gas to achieve atmospheric pressure. The alloy material within the quartz tube is then induced to melt, transforming the alloy master ingot into a liquid state. Once the pressure in the storage cylinder reaches 8 MPa, the vent valve is opened to initiate powder production.

The average diameter of powder particles was determined using a linear fitting method. Specifically, a SEM image was randomly selected, and one or multiple lines were drawn across the image to intersect as many powder particles as possible. The number of intersection points between the line(s) and particle boundaries, along with the total length of the line(s), were measured. To ensure accurate measurement of grain size, five SEM images were randomly selected for statistical analysis. The resulting data were compiled into histograms, followed by Gaussian fitting to generate a fitting curve. The final average grain size was then calculated based on the fitting data.

The laser cladding tests were conducted using RFL-A2000D semiconductor fiber-coupled laser cladding equipment (Raycus, Wuhan, China) with the Q235 alloy as the substrate. It is mainly composed of a PLC control cabinet, Yaskawa six-axis mechanical arm, Swiss Raytools coaxial powder feeding laser melting head, powder feeding device, cooling machine, voltage regulator, et al. Before cladding, the surface underwent oil removal, rust removal, and rough activation treatments. The resulting HEA coatings were designated as Si0, Si4, and Si8. To ensure a dense texture and good surface formability of the cladding layer, appropriate laser cladding parameters were selected. Key process parameters included laser power (1.5 kW), cladding speed (6 mm/s), spot diameter (3.5 mm), bonding rate (50%), powder feeding speed (10 g/min), and powder feeding mode (coaxial feed). The specific parameters used in this study are listed in Table 1. During the cladding test, the cladding direction along the Y-axis had a moving distance of 30 mm, while the X-axis had a moving distance of 1.75 mm. The resulting cladding layer measured 30 mm in length and 10 mm in width.

2.2. Analysis and Characterization Methods

The microstructure of the laser-cladded layers was analyzed using an OLYMPUS OLS4100 metallographic microscope (Shinjuku City, Tokyo). X-ray diffraction (XRD: Bruker, Billerica, MA, USA, D8 Advance) was used to examine the Si0, Si4, and Si8 MEA powders and cladding layers. The cross-section hardness of alloy coating was tested by a microhardness tester (Huayin, Xinxiang, China, HV-1000A) with an experimental load of 100 gf and load retention time of 10 s. The experimental load was 100 gf and the load retention time was 10 s. The corrosion resistance of the HEA cladding layers in 3.5 wt.% NaCl solution was tested and analyzed using an electrochemical workstation (Shanghai Chenhua, Shanghai, China, CHI660E). In the electrochemical tests, a silver chloride electrode, the sample surface, and a platinum electrode were used as the reference electrode, working electrode, and auxiliary electrode, respectively. The scanning potential ranged from −0.6 to 0.6 V, with a scanning rate of 1 mV/s. Electrochemical impedance spectroscopy tests were conducted at frequencies ranging from 100 kHz to 100 MHz. The morphology of the coating surface and cross-section was observed using a dual-beam scanning electron microscope (SEM, ThermoFisher, Waltham, MA, USA, Helios G4 CX) equipped with an energy-dispersive spectrometer (EDS) and electron backscatter diffraction (EBSD). The surface potential of the coating was analyzed using a Kelvin probe force microscope (KPFM, Bruker: Dimension Icon) with a PFQNE-AL probe. Two types of images were obtained during the Kelvin probe analysis: sample topography and the corresponding potential image. The resulting surface potential (V_{CPD}) is defined as the difference between the probe potential (V_{Tip}) and the sample potential (V_{Sample}): $V_{CPD} = V_{Sample} - V_{Tip}$. The sample potential is inversely proportional to the sample's work function. Lower potential values correspond to a higher contact potential, greater work function, and lower surface activity.

3. Results and Discussion

3.1. Microstructure and Phase Analysis of the Powder

The SEM morphologies of Si0, Si4, and Si8 HEA powders are shown in Figure 1. The powders exhibit good sphericity, with only a small proportion of dumbbell-shaped and irregularly shaped particles. Such morphology can reduce friction resistance between particles and enhance powder fluidity, facilitating uniform delivery during the laser cladding process. The average grain sizes of Si0, Si4, and Si8 powders are 28 μm, 19 μm, and 17 μm, respectively. This indicates that the addition of Si improves alloy fluidity and results in smaller particle sizes in the high-entropy alloy powders.

As shown in Figure 2, single FCC solid solution phases are exclusively observed in alloys with molar Si ratios of zero and four. However, with increased Si content, a BCC phase is gradually formed in the Si8 high-entropy power. This suggests that the addition of Si promotes the formation of the BCC phase in the alloy system [33].

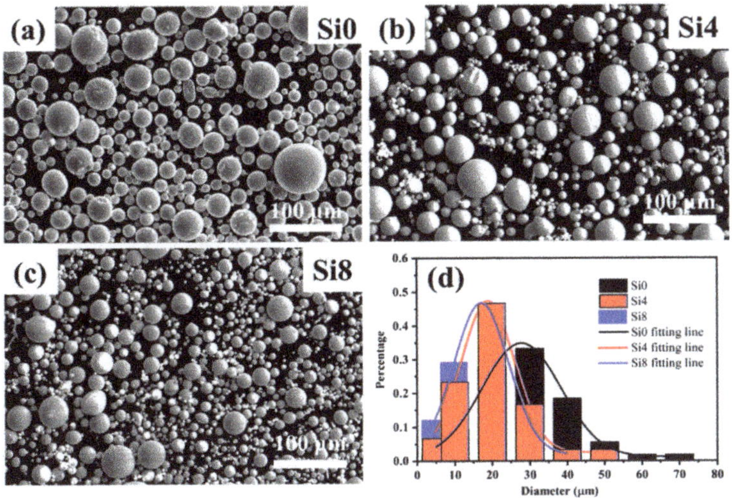

Figure 1. SEM images of (**a**) Si0, (**b**) Si4, and (**c**) Si8 high-entropy powders, along with (**d**) grain size distribution.

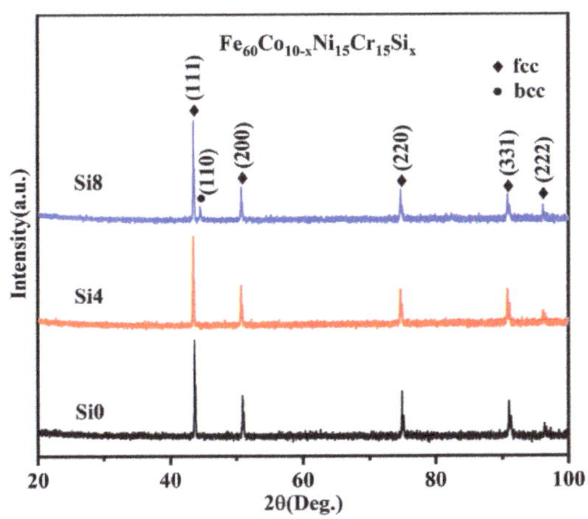

Figure 2. XRD patterns of Si0, Si4, and Si8 powders.

3.2. Microstructure and Structural Analysis of the Coating

Figure 3a shows the cross-sectional macroscopic morphology of the one-way laser-cladded sample. The sample can be divided into three regions: the coating, the heat-affected zone (HAZ), and the substrate. The coating's arched surface is positioned above the substrate, with a clearly visible fusion line between the coating and the HAZ, indicating a good metallurgical bond. The coating is dense and free of defects such as cracks and pores. Variations in dilution rates are evident, and the dilution rate (η) can be calculated using the following formula [34,35]:

$$\eta = \frac{S_1}{S_1 + S_2} \times 100\% \tag{1}$$

where S_1 is the matrix melting area, and S_2 is the coating area above the matrix surface. The coating cross-section can be modeled as two ideal arcs with different radii, as shown in Figure 3a. Based on the measured values of the thickness of the cladding layer (H_1), the thickness of the laser penetration layer (H_2), and the width of the coating (W) in Figure 3b–d, S_1 and S_2 can be calculated:

$$S_1 = \left\{ \frac{\left[\left(\frac{W}{2}\right)^2 + H_1^2\right]}{2H_1} \right\}^2 \times \left(\frac{2\pi}{360}\right) Sin^{-1} \left\{ \frac{W \cdot H_1}{\left[H_1^2 + \left(\frac{W}{2}\right)^2\right]} \right\} - \frac{W\left[\left(\frac{W}{2}\right)^2 - H_1^2\right]}{4H_1} \quad (2)$$

$$S_1 = \left\{ \frac{\left[\left(\frac{W}{2}\right)^2 + H_2^2\right]}{2H_2} \right\}^2 \times \left(\frac{2\pi}{360}\right) Sin^{-1} \left\{ \frac{W \cdot H_2}{\left[H_2^2 + \left(\frac{W}{2}\right)^2\right]} \right\} - \frac{W\left[\left(\frac{W}{2}\right)^2 - H_2^2\right]}{4H_2} \quad (3)$$

The thickness of the coatings is about 250 μm, with no observable defects such as macroscopic cracks, porosity, or unmelted powder in the cladding zone. As shown in Figure 3b–d, the HEA coating can be divided into four regions along the cross-section: coating, HAZ, transition zone (TZ), and substrate. The transition zone confirms the formation of a good metallurgical bond between the alloy coating and substrate using laser cladding technology. As the Si content increases, the dilution rate also increases (28.8% for Si0, 31.5% for Si4, and 34.4% for Si8). The dilution rate primarily depends on the energy absorbed by the substrate, which is closely related to the energy absorbed by the prior layer. Since the process parameters, as well as the thickness, density, and surface roughness of the cladding layer, are nearly identical in this study, the absorbed energy is mainly influenced by the chemical composition of the cladding layer. The change in dilution rate with Si content results from the interaction of the above factors. The experimental results suggest that high absorptivity compensates for the adverse effect of increased energy on substrate melting. Specifically, when x is increased to four, the high absorptivity due to increased Si content outweighs the increase in dilution rate.

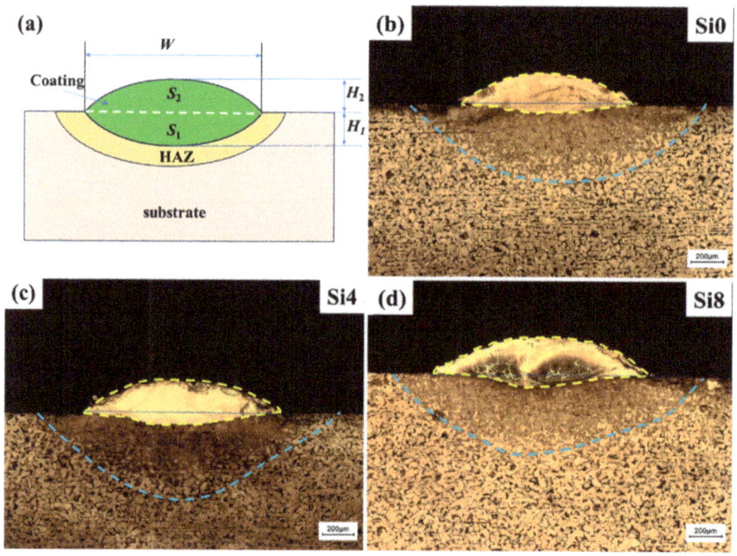

Figure 3. (**a**) Simplified profile of the coating and optical cross-sectional profiles of (**b**) Si0, (**c**) Si4, and (**d**) Si8 HEA coatings with a one-way path.

Si was added to the FeCoCrNi HEA system to maintain its single-phase FCC solid solution structure without forming secondary or precipitate phases. A single-phase FCC solid solution can only exist in the FeCoCrNi system when the Si content is below 5.88 at.% [24]. Figure 4 shows the XRD pattern of the HEA coatings. The results confirm an FCC structure after laser cladding, with no evident BCC peaks, indicating that the coatings derived from the three alloy powders are FCC structure. During the formation of high-entropy alloys, the phase development is closely associated with the cooling rate. In general, a high cooling rate facilitates the promotion of martensitic transformation [36]. The discrepancies in the XRD results between the powder (see Figure 2) and the coating are similar to results of our previous studies [37]. This suggests that the cooling rate exerts a substantial influence on the microstructure of FeCoCrNiSi high-entropy alloys during the preparation process.

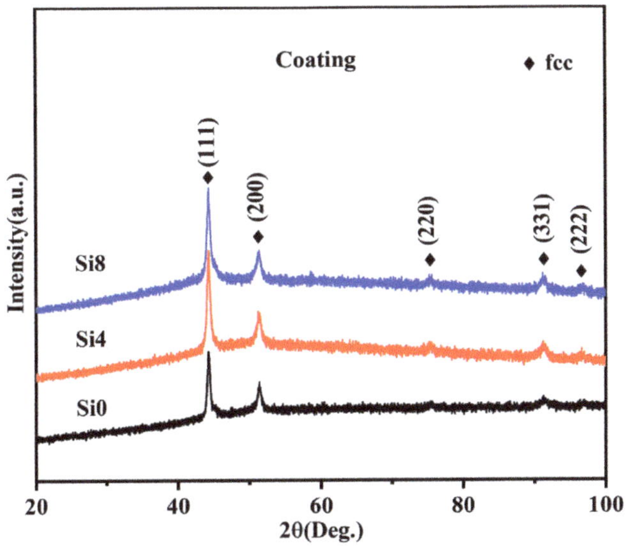

Figure 4. XRD patterns of Si0, Si4, and Si8 high-entropy coatings.

To further analyze the microstructure and grain orientation of the cladding layer after LC, EBSD analysis was performed on the surface of Si0, Si4, and Si8 MEA coatings. As shown in Figure 5a–c, the surface structures of all coatings exhibit FCC symmetry. Additionally, the grain sizes of the Si0, Si4, and Si8 coatings were measured to be 98.1 μm, 38.5 μm, and 34.8 μm, respectively. The addition of Si enhances grain growth potential in the high-entropy coating. As grains grow, the number of grain boundaries decreases, reducing grain boundary corrosion during the corrosion process and improving the coating's corrosion resistance.

The cross-sectional morphologies from the coating to the substrate for Si0, Si4, and Si8 coatings are shown in Figure 6, revealing an approximate coating thickness of 350 μm (Figure 6a–c). No visible cracks, pores, or other defects were observed, which is consistent with the results of Figure 3 and further confirms the high quality of the prepared coatings. The TZ width is approximately 10 μm (Figure 6), indicating the formation of a metallurgical bond between the MEA coating and the Q235 steel substrate via laser cladding. EDS line scans in Figure 6d–f confirm a compact bonding interface between the coating and substrate. Near the interface, planar crystal structures transition into cellular crystals as one moves toward the center of the melting pool. Element distributions along the EDS scan line, indicated by the red arrow, are shown in Figure 6d–f. Iron (Fe) content is higher near the interface (at a distance of 45–50 μm), likely due to the elevated temperature during laser cladding, which facilitates Fe diffusion from the Q235 steel into the coating. Minor amounts of Ni, Cr, and Si were detected in the substrate, indicating slight diffusion from the coating into the substrate. This interdiffusion further confirms the strong metallurgical bond formed between the MEA coating and the Q235 steel substrate.

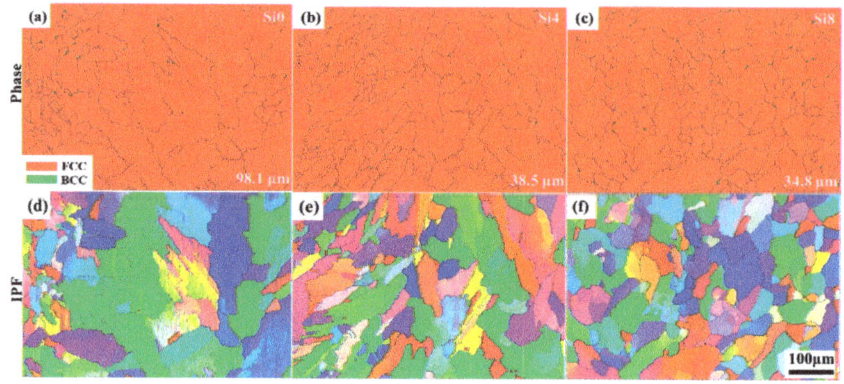

Figure 5. Phase map and inverse pole figure of the surface of (**a**,**d**) Si0, (**b**,**e**) Si4, and (**c**,**f**) Si8 HEA coatings; the average grain size is shown in (**a**–**c**).

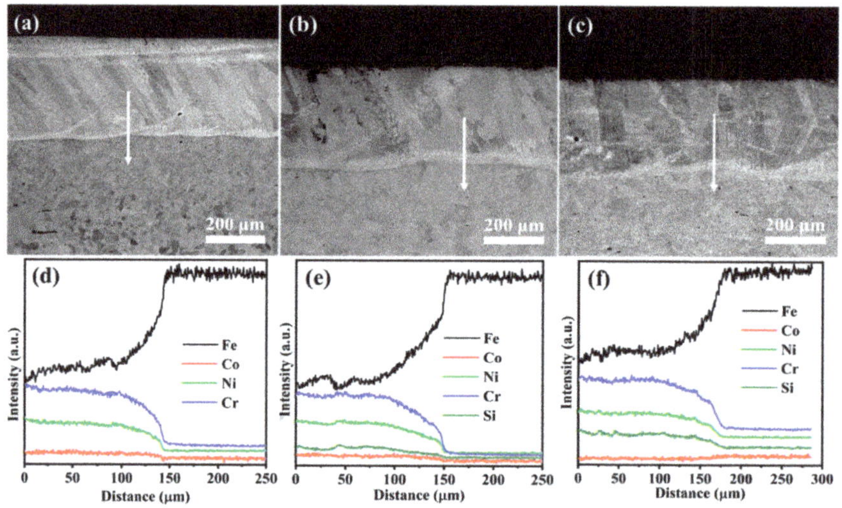

Figure 6. Cross-sectional SEM images and energy-dispersive X-ray spectroscopy results for (**a**,**d**) Si0, (**b**,**e**) Si4, and (**c**,**f**) Si8 high-entropy alloy coatings.

The layered structure formed after laser cladding plays a crucial role in the fatigue resistance and strength of the cladding layer. Phase images and inverse pole figures from EBSD analysis of the cross-sections of Si0, Si4, and Si8 coatings are shown in Figure 7. The cross-sections of the Si0, Si4, and Si8 coatings exhibit a dense, well-bonded structure, characteristic of a typical metallurgical interface. The lower region of the cross-section primarily consists of coarse grains, while fine grains are observed at the joint regions. Coarse grains reappear near the transition zones. For the Si4 coating, a small amount of the BCC phase is precipitated at the boundaries of the coarse grains. In the Si8 coating, the BCC structure is more prominently observed within both the grain boundaries and interiors. This indicates that increasing the Si content enhances the powder's fluidity during melting, leading to improved bonding with the substrate.

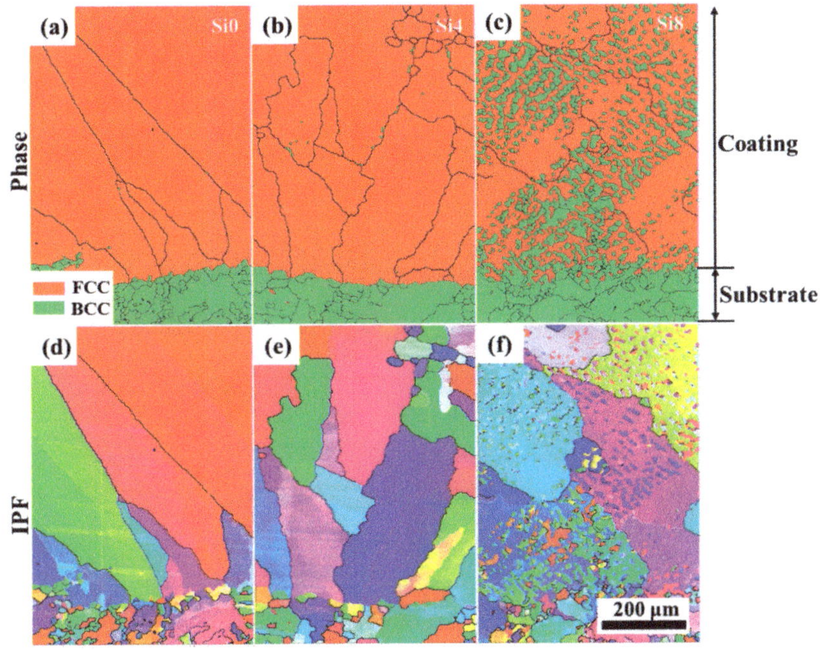

Figure 7. Phase images and inverse pole figures from EBSD analysis of the cross-sections of Si0, Si4, and Si8 coatings, phase of (**a**) Si0, (**b**) Si4 and (**c**) Si8 coatings, IPF of (**d**) Si0, (**e**) Si4 and (**f**) Si8 coatings.

3.3. Microhardness

To elucidate the mechanical behavior of the HEA coating, the hardness variation in Si0, Si4, and Si8 coatings across the cross-sectional area from the surface to the substrate is illustrated in Figure 8. It is evident that the average hardness values for the coatings are 229 Hv (Si0), 259 Hv (Si4), and 285 Hv (Si8), which are significantly higher than those of the substrate (162 Hv). The elevated hardness of the coating can be attributed to several factors: firstly, the relatively small atomic radius of silicon (Si) results in a minimal BCC phase when its content is low ($\leq$4), with the coating's phase structure primarily being FCC. As the Si content increases (up to Si8), Si atoms dissolve into the interstitial sites of the FCC phase, leading to an increase in both the BCC phase content and solid solubility. The incorporation of Si partially displaces other metal elements, and the difference in atomic radii enhances lattice distortion, thereby significantly increasing the solid solution strengthening effect and, consequently, the hardness. Notably, the hardness of the coating near the heat-affected zone (HAZ) increases due to the substantial formation of the BCC phase upon fusion with the substrate in the HAZ, where the hardness of the BCC phase surpasses that of the FCC phase, resulting in a gradual increase in hardness from the surface to the bonding zone. Additionally, the grain refinement induced by the rapid solidification during laser cladding, as demonstrated in Figure 5, contributes to this phenomenon.

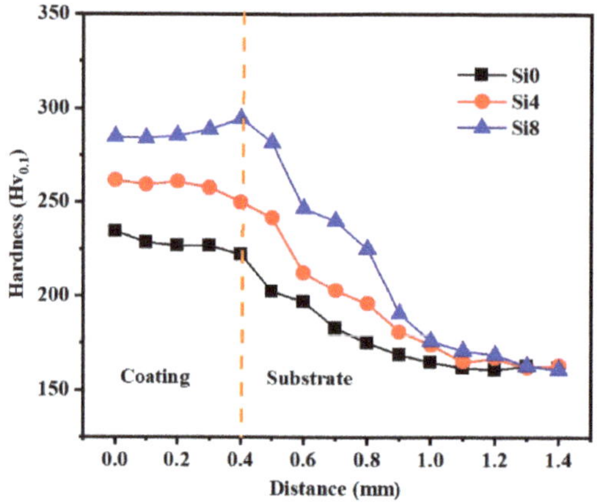

Figure 8. Variations in the hardness of Si0, Si4 and Si8 coatings along an increasing distance from the coating to the Q235 steel substrate.

3.4. Corrosion Behaviors

Figure 9 illustrates the polarization curves of the Si0, Si4, and Si8 coating surfaces in a 3.5 wt.% NaCl solution. The self-corrosion potentials of Si0, Si4, and Si8 coatings are −148 mV, −157 mV, and −0.347 mV, respectively. While the self-corrosion potentials of Si0 and Si4 coatings are relatively close, the increase in Si content to eight leads to a sharp decrease in the self-corrosion potential. The Si0 coating exhibits a zigzag fluctuation in the passivation region, which is attributed to the breakdown and re-passivation of the passive film, causing pitting corrosion [38,39]. In contrast, Si8 does not display passivation in the NaCl solution, indicating poor corrosion resistance. However, Si0 and Si4 coatings show clear passivation regions with high pitting potentials, highlighting the improved corrosion resistance due to Si addition [40]. The reduced corrosion resistance of the Si8 coating may be due to Si enrichment at grain boundaries, potentially reducing intergranular corrosion of the coating. It is noted that the Si addition can cause grain refinement and increase the number of grain boundaries, as well as introduce defects during crystallization [33,41,42]. Therefore, with the addition of Si elements, the grain size of the Si8 coating decreases obviously and thus a more pronounced enrichment may occur, finally leading to decreased corrosion resistance. When the Si content in the alloy is low, the Si4 coating does not achieve the necessary concentration to mitigate corrosion, and thus the corrosion performance of the alloy remains unaffected. Consequently, the grain boundaries become preferential sites for corrosion, resulting in the poorest corrosion performance for the Si8 coating.

Due to the similar corrosion resistance of Si0 and Si4 coatings, surface potential analysis was performed. The surface morphology and contact potential (V_{CPD}) of Si0 and Si4 coatings, characterized using Kelvin probe force microscopy, are shown in Figure 10. Results indicate that lower V_{CPD} corresponds to a higher work function, reduced surface activity, and greater corrosion resistance [43]. Figure 10a,c reveal that as Si content increases, the coating surface transitions from dendritic to equiaxed crystals, with a reduction in grain size [42,44]. Figure 10b,d show minimal potential differences between grains and grain boundaries, as well as between strip and dendritic crystals on the surfaces of Si4 and Si8 coatings. XRD and EBSD results confirm that the coating surfaces predominantly consist of a single FCC phase, resulting in a uniform alloy composition. This uniformity minimizes potential differences and reduces the influence of alloy composition on the corrosion process.

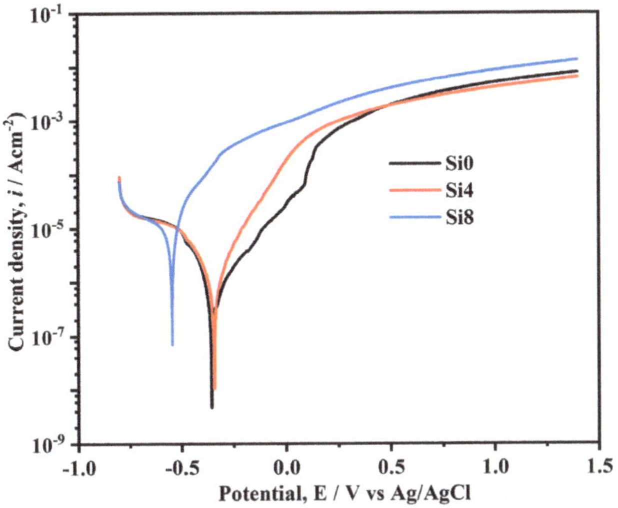

Figure 9. Tafel curves of Si0, Si4, and Si8 coatings in 3.5 wt.% NaCl solution.

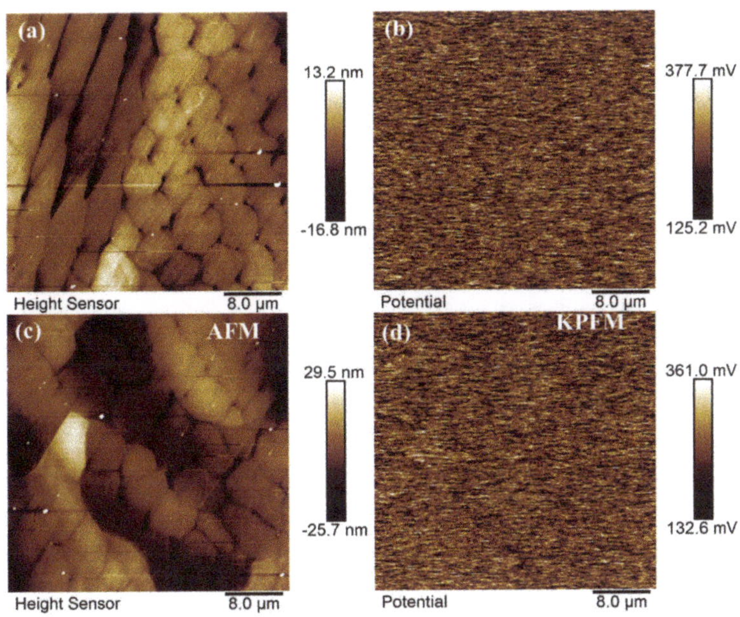

Figure 10. Surface morphology and corresponding KPFM images of (**a**,**b**) Si0 and (**c**,**d**) Si4 coatings.

To further clarify the differences in corrosion behavior between Si0 and Si4 coatings, post-corrosion morphology and the corresponding line scan results are presented in Figure 11. Both coatings exhibit equiaxed crystal structures, with Si4 having smaller grains. After corrosion, small pitting pits appear at the grain boundaries of both coatings. Elemental analysis (Figure 11b,d) shows a reduction in Fe content and an enrichment of Cr and Ni elements at the grain boundaries of the Si0 alloy. In the Si4 coating, Si, Cr, and Ni element are enriched at the grain boundaries, suggesting that Si addition promotes Cr and Ni segregation in this region. Corrosion typically initiates at grain boundaries, where Cr and Si form oxide films such as Cr_2O_3 and SiO_2. As these oxides grow, thermal stresses can cause cracking and spallation, exposing fresh metal [41]. However, the presence of SiO_2 in the passivation film may enhance the stability of Cr_2O_3, improving the corrosion resistance of the Si4 coating [30].

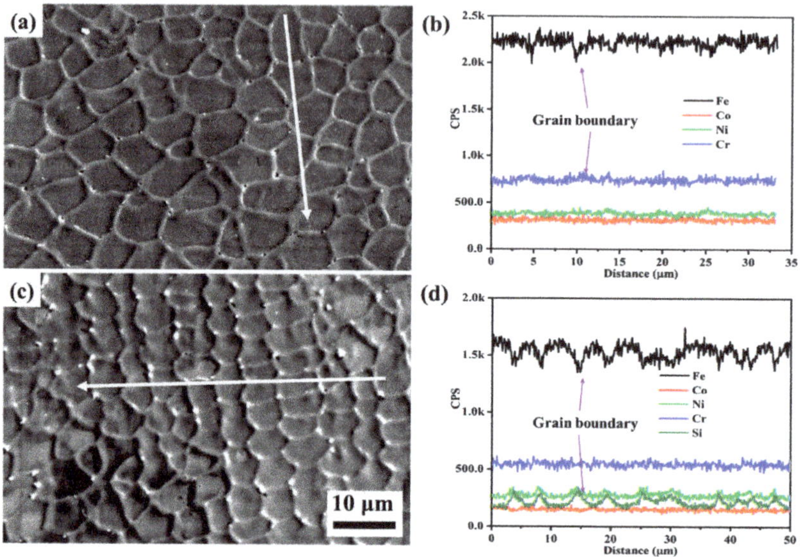

Figure 11. Secondary electron images and line scan results of (**a**,**b**) Si0 and (**c**,**d**) Si4 coatings after corrosion tests in 3.5 wt.% NaCl solution.

4. Conclusions

This study explored the effect of Si addition on the microstructure and corrosion resistance of $Fe_{60}Co_{10-x}Ni_{15}Cr_{15}Si_x$ (x = 0, 4, and 8) high-entropy alloy coatings. High-entropy alloy powder was synthesized via aerosol technology, and coatings were prepared using laser cladding. The main findings are as follows:

1. Si addition enhances the powder's fluidity during melting, leading to improved bonding with the substrate.
2. Si addition leads to its enrichment at grain boundaries, hindering grain growth and refining the grain size of the coating.
3. For the studied coatings, all samples exhibit an FCC structure, with only minor BCC phases at grain boundaries. At the coating–substrate interface, increasing Si content results in a transition from a single-phase FCC to an FCC + BCC structure.
4. The Si4 coating demonstrates superior corrosion resistance, attributed to the Si-enhanced enrichment of corrosion-resistant elements at grain boundaries. This enrichment stabilizes the passivation film, effectively improving corrosion resistance while maintaining low production costs.

Author Contributions: Investigation, D.W.; Methodology, C.H.; Investigations, Z.D.; Resources, Investigation, W.L.; Formal analysis, Writing—review & editing, S.Z.; Investigation, Formal analysis, F.L.; Investigation, J.L. All authors have read and agreed to the published version of the manuscript.

Funding: This research was funded by Doctoral Research Start-up Fund of Henan Open University grant number BSJH-2024-08, National Natural Science Foundation of China grant number 52101203, Education Department of Henan Province grant number 23A430014, and Nuclear Material Technology Innovation Fund for National Defense Technology Industry grant number ICNM-2023-YZ-02.

Institutional Review Board Statement: Not applicable.

Informed Consent Statement: Not applicable.

Data Availability Statement: The original contributions presented in this study are included in the article. Further inquiries can be directed to the corresponding authors.

Acknowledgments: The authors gratefully acknowledge the Doctoral Research Start-up Fund of Henan Open University grant number BSJH-2024-08, National Natural Science Foundation of China grant number 52101203, Education Department of Henan Province grant number 23A430014, and Nuclear Material Technology Innovation Fund for National Defense Technology Industry grant number ICNM-2023-YZ-02.

Conflicts of Interest: Author Wenqiang Li is employed by the company Henan Coal Science Academy Wear-resistant Technology Co., Ltd. The remaining authors declare that the research was conducted in the absence of any commercial or financial relationships that could be construed as a potential conflict of interest.

References

1. George, E.P.; Raabe, D.; Ritchie, R.O. High-entropy alloys. *Nat. Rev. Mater.* **2019**, *4*, 515–534. [CrossRef]
2. Li, W.D.; Xie, D.; Li, D.Y.; Zhang, Y.; Gao, Y.F.; Liaw, P.K. Mechanical behavior of high-entropy alloys. *Prog. Mater. Sci.* **2021**, *118*, 100777. [CrossRef]
3. Liu, C.; Gao, Y.; Chong, K.; Guo, F.Q.; Wu, D.T.; Zou, Y. Effect of Nb content on the microstructure and corrosion resistance of FeCoCrNiNbx high-entropy alloys in chloride ion environment. *J. Alloys Compd.* **2023**, *935*, 168013. [CrossRef]
4. Yin, B.L.; Maresca, F.; Curtin, W.A. Vanadium is an optimal element for strengthening in both fcc and bcc high-entropy alloys. *Acta Mater.* **2020**, *188*, 486–491. [CrossRef]
5. Xiao, D.H.; Zhou, P.F.; Wu, W.Q.; Diao, H.Y.; Gao, M.C.; Song, M.; Liaw, P.K. Microstructure, mechanical and corrosion behaviors of AlCoCuFeNi-(Cr,Ti) high entropy alloys. *Mater. Des.* **2017**, *116*, 438–447. [CrossRef]
6. Shi, Z.L.; Liu, Y.Z.; Zhang, H.R.; Li, C.Z.; Chen, S.N.; Yang, Y.J.; Liang, S.X.; Ma, M.Z. Synergistic strength-ductility enhancement of CoCrFeNi high-entropy alloys with regulated Co/Cr atomic ratios. *Mat. Sci. Eng. A* **2024**, *912*, 146995. [CrossRef]
7. Zhu, L.D.; Xue, P.S.; Lan, Q.; Meng, G.; Ren, Y.; Yang, Z.C.; Xu, P.H.; Liu, Z. Recent research and development status of laser cladding: A review. *Opt. Laser Technol.* **2021**, *138*, 106915. [CrossRef]
8. Liang, Y.; Liao, Z.; Zhang, L.; Cai, M.; Wei, X.; Shen, J. A review on coatings deposited by extreme high-speed laser cladding: Processes, materials, and properties. *Opt. Laser Technol.* **2023**, *164*, 109472. [CrossRef]
9. Arif, Z.U.; Khalid, M.Y.; Rehman, E.U.; Ullah, S.; Atif, M.; Tariq, A. A review on laser cladding of high-entropy alloys, their recent trends and potential applications. *J. Manuf. Process* **2021**, *68*, 225–273. [CrossRef]
10. Liu, H.; Yang, D.; Jiang, Q.; Jiang, Y.; Yang, W.; Liu, L.; Zhang, L.-C. Additive manufacturing of metallic glasses and high-entropy alloys: Significance, unsettled issues, and future directions. *J. Mater. Sci. Technol.* **2023**, *140*, 79–120. [CrossRef]
11. Xiao, J.K.; Tan, H.; Wu, Y.Q.; Chen, J.; Zhang, C. Microstructure and wear behavior of FeCoNiCrMn high entropy alloy coating deposited by plasma spraying. *Surf. Coat. Technol.* **2020**, *385*, 125430. [CrossRef]
12. Zendejas, M.L.; Riekehr, L.; Jansson, U. Phase formation in magnetron sputtered CrMnFeCoNi high entropy alloy. *Surf. Coat. Technol.* **2020**, *403*, 126323. [CrossRef]
13. Liu, Z.; Liu, G.; Miao, X.; Sun, L.; Han, Z.; Zhang, G. Microstructure and wear-resistance of the CoCrFeNiMo high entropy alloy coatings by modulated arc currents using plasma transferred arc weld technique. *Intermetallics* **2024**, *175*, 108506. [CrossRef]
14. Tanigawa, D.; Abe, N.; Tsukamoto, M.; Hayashi, Y.; Yamazaki, H.; Tatsumi, Y.; Yoneyama, M. The effect of particle size on the heat affected zone during laser cladding of Ni–Cr–Si–B alloy on C45 carbon steel. *Opt. Laser Eng.* **2018**, *101*, 23–27. [CrossRef]
15. Li, J.C.; Huang, Y.X.; Meng, X.C.; Xie, Y.M. A review on high entropy alloys coatings: Fabrication processes and property assessment. *Adv. Eng. Mater.* **2019**, *21*, 1900343. [CrossRef]
16. Bae, J.W.; Seol, J.B.; Moon, J.; Sohn, S.S.; Jang, M.J.; Um, H.Y.; Lee, B.J.; Kim, H.S. Exceptional phase-transformation strengthening of ferrous medium-entropy alloys at cryogenic temperatures. *Acta Mater.* **2018**, *161*, 388–399. [CrossRef]
17. Zhang, K.S.; Zhang, X.H.; Zhang, E.; Wei, R.; Wang, L.L.; Chen, J.L.; Yuan, S.H.; Han, Z.H.; Chen, C.; Li, F.S. Strengthening of ferrous medium entropy alloys by promoting phase transformation. *Intermetallics* **2021**, *136*, 107265. [CrossRef]
18. Han, Z.H.; Li, J.Z.; Tian, Y.; Tian, A.L.; Sun, A.R.; Wei, R.; Liu, G. Annealing-induced abnormally ultrahigh cryogenic strength-ductility synergy in a cold rolled ferrous medium entropy alloy. *Intermetallics* **2023**, *159*, 107933. [CrossRef]

19. Dong, M.L.; Liu, P.W.; Wang, C.H.; Wang, Y.H.; Tang, X.Y.; He, M.X.; Liu, J.Q. Microstructure and properties of FeCoNiCr and FeCoNiCrW high entropy alloy coatings by electro-deposition. *Intermetallics* **2024**, *175*, 108492. [CrossRef]
20. Rong, Z.Y.; Wang, C.H.; Wang, Y.; Dong, M.L.; You, Y.; Wang, J.N.; Liu, H.N.; Liu, J.Q.; Wang, Y.H.; Zhu, Z.Y. Microstructure and properties of FeCoNiCrX (X Mn, Al) high-entropy alloy coatings. *J. Alloys Compd.* **2022**, *921*, 166061. [CrossRef]
21. Kong, L.C.; Shi, C.W.; Hao, X.; Wang, S.H.; Huo, Y.S.; Zhu, Z.H.; Sun, J.H. Effect of Ti and Mo content changes on microstructure and properties of laser cladding FeCoCrNiMn high entropy alloy coatings. *Intermetallics* **2024**, *175*, 108509. [CrossRef]
22. Chaudhary, V.; Soni, V.; Gwalani, B.; Ramanujan, R.V.; Banerjee, R. Influence of non-magnetic Cu on enhancing the low temperature magnetic properties and Curie temperature of FeCoNiCrCu(x) high entropy alloys. *Scripta Mater.* **2020**, *182*, 99–103. [CrossRef]
23. Xu, Z.K.; Hou, Y.H.; Li, G.Q. Atomic-scale investigation of the synergistic impact of Mo and Nb addition on enhancing the performance of laser cladding FeCoCrNi-based high entropy alloy coatings. *Surf. Coat. Technol.* **2024**, *490*, 131162. [CrossRef]
24. Lin, D.Y.; Xu, L.Y.; Li, X.J.; Jing, H.; Qin, G.; Pang, H.; Minami, F. A Si-containing FeCoCrNi high-entropy alloy with high strength and ductility synthesized in situ via selective laser melting. *Addit. Manuf.* **2020**, *35*, 101340. [CrossRef]
25. Cemin, F.; Jimenez, M.; Leidens, L.M.; Figueroa, C.A.; Alvarez, F. A thermodynamic study on phase formation and thermal stability of AlSiTaTiZr high-entropy alloy thin films. *J. Alloys Compd.* **2020**, *838*, 155580. [CrossRef]
26. Moravcik, I.; Cizek, J.; Zapletal, j.; Kovacova, Z.; Vesely, J.; Minarik, P.; Kitzmantel, M.; Neubauer, E.; Dlouhy, I. Microstructure and mechanical properties of $Ni_{1.5}Co_{1.5}CrFeTi_{0.5}$ high entropy alloy fabricated by mechanical alloying and spark plasma sintering. *Mater. Des.* **2017**, *119*, 141–150. [CrossRef]
27. Yang, X.; Zhang, Y. Prediction of high-entropy stabilized solid-solution in multi-component alloys. *Mater. Chem. Phys.* **2012**, *132*, 233–238. [CrossRef]
28. Kumar, J.; Linda, A.; Sadhasivam, M.; Pradeep, K.G.; Gurao, N.P.; Biswas, K. The effect of Si addition on the structure and mechanical properties of equiatomic CoCrFeMnNi high entropy alloy by experiment and simulation. *Materialia* **2023**, *27*, 101707. [CrossRef]
29. Wei, D.X.; Gong, W.; Tsuru, T.; Lobzenko, I.; Li, X.; Harjo, S.; Kawasaki, T.; Do, H.-S.; Bae, J.W.; Wagner, C.; et al. Si-addition contributes to overcoming the strength-ductility trade-off in high-entropy alloys. *Int. J. Plast.* **2022**, *159*, 103443. [CrossRef]
30. Nagarjuna, C.; Lee, H.; Dewangan, S.K.; Rao, K.P.; Pillai, G.M.; Kumar, V.; Ahn, B. Understanding the role of Si alloying on the structural, mechanical, wear and high temperature oxidation behavior of CrFeNiTiX (X=Si) high entropy alloys. *J. Mater. Res. Technol.* **2024**, *33*, 5119–5135. [CrossRef]
31. Huang, L.; Wang, X.J.; Jia, F.C.; Zhao, X.; Huang, B.; Ma, J.; Wang, C. Effect of Si element on phase transformation and mechanical properties for FeCoCrNiSix high entropy alloys. *Mater. Lett.* **2021**, *282*, 128809. [CrossRef]
32. Gao, Q.Y.; Feng, S.L.; Wei, R.; Chen, W.; Zhu, W.; Li, F. A novel low-cost and ultrahigh strength FeCoNiCrSi ferrous medium-entropy alloy at room and liquid nitrogen temperatures. *Mater. Lett.* **2022**, *328*, 133119. [CrossRef]
33. Lin, T.X.; Feng, M.Y.; Lian, G.F.; Lu, H.; Chen, C.; Huang, X. Effects of Si content on the microstructure and properties of CoCrFeMnNiSix high-entropy alloy coatings by laser cladding. *Mater. Charact.* **2024**, *216*, 114246. [CrossRef]
34. Liu, Y.Y.; Yao, Z.P.; Zhang, P.; Xu, Z.; Lin, S.; He, M.; Lu, S.; Wu, X. Tailoring high-temperature oxidation resistance of FeCrMnVSi high entropy alloy coatings via building Si-rich dendrite microstructure. *Appl. Surf. Sci.* **2022**, *606*, 154862. [CrossRef]
35. Cai, Y.C.; Chen, Y.; Manladan, S.M.; Luo, Z.; Gao, F.; Li, L. Influence of dilution rate on the microstructure and properties of FeCrCoNi high-entropy alloy coating. *Mater. Des.* **2018**, *142*, 124–137. [CrossRef]
36. Mao, C.L.; Liu, C.X.; Yu, L.M.; Li, H.; Liu, Y. Discontinuous lath martensite transformation and its relationship with annealing twin of parent austenite and cooling rate in low carbon RAFM steel. *Mater. Des.* **2021**, *197*, 109252. [CrossRef]
37. Yuan, S.H.; Wu, S.J.; Chen, T.T.; Fu, Q.; Wei, R.; Chen, C.; Wang, T.; Cai, Y.; Li, F. Microstructure and corrosion behavior of Co-free FeCrNiSi0.4 medium entropy alloy coating fabricated by laser cladding. *Intermetallics* **2023**, *162*, 108024. [CrossRef]
38. Yang, H.; Liu, X.S.; Li, A.X.; Li, R.; Xu, S.; Zhang, M.; Yu, P.; Yu, S.; Jiang, M.; Huo, C.; et al. Effect of silicon addition on the corrosion resistance of Al0.2CoCrFe1.5Ni high-entropy alloy in saline solution. *J. Alloys Compd.* **2023**, *964*, 171226. [CrossRef]
39. Zhang, M.D.; Shi, X.L.; Li, Z.Y.; Xu, H.; Li, G. Corrosion behaviors and mechanism of $CrFeNi_2$ based high-entropy alloys. *Corros. Sci.* **2022**, *207*, 110562. [CrossRef]
40. Yu, K.P.; Feng, S.H.; Ding, C.; Yu, P.; Huang, M.X. Improving anti-corrosion properties of CoCrFeMnNi high entropy alloy by introducing Si into nonmetallic inclusions. *Corros. Sci.* **2022**, *208*, 110616. [CrossRef]
41. Liu, H.; Sun, S.F.; Zhang, T.; Zhang, G.; Yang, H.; Hao, J. Effect of Si addition on microstructure and wear behavior of AlCoCrFeNi high-entropy alloy coatings prepared by laser cladding. *Surf. Coat. Technol.* **2021**, *405*, 126522. [CrossRef]
42. Zhu, Z.X.; Liu, X.B.; Liu, Y.F.; Zhang, S.-Y.; Meng, Y.; Zhou, H.-B.; Zhang, S.-H. Effects of Cu/Si on the microstructure and tribological properties of FeCoCrNi high entropy alloy coating by laser cladding. *Wear* **2023**, *512–513*, 204533. [CrossRef]

43. Shi, Z.J.; Wang, Z.B.; Wang, X.D.; Zhang, S.; Zheng, Y.G. Effect of thermally induced B2 phase on the corrosion behavior of an Al0.3CoCrFeNi high entropy alloy. *J. Alloys Compd.* **2022**, *903*, 163886. [CrossRef]
44. Zhang, S.L.; Di, J.; Sun, S.D.; Zhang, B. Microstructure and tribological properties of FeCrCoMnSix high-entropy alloy coatings. *Coatings* **2024**, *14*, 1476. [CrossRef]

materials

MDPI

Article

Protective Properties of Silane Composite Coatings Modified with Poly(3,4-ethylenedioxythiophene) with Heteropolyacid on X20Cr13 and 41Cr4 Steel

Aleksandra Kucharczyk-Kotlewska [1], Lidia Adamczyk [1], Krzysztof Miecznikowski [2] and Agata Dudek [1,*]

1 Department of Materials Engineering, Faculty of Production Engineering and Materials Technology, Czestochowa University of Technology, Aleja Armii Krajowej 19, 42-200 Czestochowa, Poland; aleksandra.kucharczyk@pcz.pl (A.K.-K.); lidia.adamczyk@pcz.pl (L.A.)

2 Faculty of Chemistry, University of Warsaw, Ludwika Pasteura 1, 02-093 Warsaw, Poland; kmiecz@chem.uw.edu.pl

* Correspondence: agata.dudek@pcz.pl

Abstract: This paper describes the methodology of the preparation and analyses of the structure and anticorrosion properties of silane coatings modified with poly(3,4-ethylenedioxythiophene) (PEDOT) with phosphododecamolybdic acid (PMo_{12}). Protective coatings, consisting of vinyltrimethoxysilane (VTMS), PEDOT powder with PMo_{12} admixture (at different concentrations), and ethanol, were deposited on X20Cr13 and 41Cr4 steels by immersion. The physicochemical properties of these silane coatings (e.g., surface morphology, thickness, roughness, and adhesion to the substrate) were elucidated using a digital microscope, a Fourier transform infrared spectrophotometer with attenuated total reflectance, and various electrochemical diagnostic techniques. Protective properties were assessed in acidified sulfate solutions with and without chloride ions (pH 2). Experimental results have shown that this coating displayed the effective protection of steel against general and pitting corrosion, stabilized the corrosion potential in the passive state, and provided barrier protection.

Keywords: vinyltrimethoxysilane (VTMS); poly(3,4-ethylenedioxythiophene) PEDOT; heteropoly-acid; phosphomolybdic acid (PMo_{12})

1. Introduction

Stainless steel has been widely used in many industries, especially those where high-quality and durable materials are critical. In the construction industry in particular, it is valued for its modern, elegant appearance and its ability to be easily formed into a variety of shapes [1–4]. With its resistance to corrosion, mechanical damage, and extreme temperatures, stainless steel is an ideal material for the manufacture of components such as pipes, balustrades, columns, or lift cabins [5–8]. Despite its favorable properties, stainless steel, including steel with more than 10.5% chromium, can rust under adverse conditions. Any steel reacts with oxygen to form an oxide layer. In conventional steels, this process leads to corrosion as oxygen reacts with the iron atoms to form a rougher surface. In stainless steel, chromium reacts with oxygen to form a passive layer that protects the material from further corrosion reactions [9–11].

Advanced protective coating technologies are being developed to increase the protection of steel against corrosion [12,13]. One such technology is silane coatings, which contain both organic (methoxy or ethoxy) groups and inorganic silicon [14–17]. Silanes were initially used as adhesion promoters for glass substrates, but their use has expanded significantly to include corrosion protection for various metals such as aluminum, iron, and steel [18–26]. Silane technology became particularly important in the 1990s when alternatives to toxic chromium (VI) coatings were sought [27–30].

In addition to silanes, conductive polymers such as poly(3,4-ethylenedioxythiophene) (PEDOT) are beginning to play an increasingly important role in corrosion protection [31–40].

Due to their conductive properties, these polymers find applications in a wide range of fields, from electronics to sensors and catalysis [41–47]. By combining silane coatings with conductive polymers such as PEDOT, hybrid coatings can be created that significantly improve the corrosion protection of steel while providing additional functions such as electrical conductivity [48–53].

PEDOT (poly(3,4-ethylenedioxythiophene)) coatings can be obtained by chemical or electrochemical polymerization of the EDOT monomer in organic solvents such as H_2SO_4 [54–57]. A study [58] has shown that the use of the micellar agent BRIJ significantly improves the degree of cross-linking and structural properties of PEDOT coatings, resulting in better protective and functional performance.

A literature review [59–63] also suggests that phosphoric acid and heteropolyacids (HPAs) play a key role in the electropolymerization of thiophenes, leading to the formation of hybrid PEDOT coatings with polyoxometalates such as phosphomolybdate and phosphovolphramate. HPAs are distinguished by their strong acidic properties [64,65] and excellent proton conductivity under ambient conditions [66–68]. Their ability to rapidly and reversibly transfer electrons makes them of particular interest as potential electrocatalysts in redox reactions [69–74]. Such hybrid composite coatings, combining the properties of PEDOT and HPA, represent a promising solution for corrosion protection and functional materials in advanced technological applications [75–81].

The aim of this study was to develop a method for the deposition of anticorrosion coatings on X20Cr13 and 41Cr4 steels using the EDOT monomer with phosphododecamolybdic acid ($H_3PMo_{12}O_{40}$) in a sol–gel solution containing VTMS silane. PEDOT was chosen for its high chemical stability, low toxicity, and ease of application, making it a suitable material for corrosion protection. In the present study, the proposed coatings displayed the effective protection of steel against general and pitting corrosion (presence of phosphododecamolybdic anions), stabilized the corrosion potential in the passive state (existence of conducting polymer—PEDOT), and provided anodic barrier protection. Furthermore, despite the extensive use of PEDOT in other areas, the use of VTMS/PEDOT/PMo_{12} coatings in the corrosion protection of X20Cr13 and 41Cr4 steels has not been reported in the literature.

2. Materials and Methods

2.1. Materials

The following reagents were used to prepare the modification solution: vinyltrimethoxysilane (VTMS 99%, Sigma Aldrich, St. Louis, MO, USA), anhydrous ethyl alcohol (EtOH, Sigma Aldrich, St. Louis, MO, USA), 3,4-ethylenedioxythiophene monomer (EDOT, Sigma Aldrich, St. Louis, MO, USA), and phosphododecamolybdic acid (PMo_{12}, Fluka, Buchs, Switzerland). All reagents had a p.a. purity grade.

Table 1 illustrates the chemical composition of the alloy steels used in the experiments. Steel specimens with a diameter of 5 mm were placed in polymethylmethacrylate holders using epoxy resin. The working area of the specimens was 0.2 cm^2. The specimens were mechanically polished on increasingly fine-grained sandpaper up to No. 2000 each time before conducting experiments, followed by washing with distilled water and ethanol. Before the application of the coating, each specimen was washed with acetone to degrease the surface.

Table 1. Chemical composition of the alloy steels used in the experiments.

	C %	Cr%	Si%	Ni%	Mn%	V%	P%	S%	Ti%	Mo%	W%
X20Cr13	0.17	12.6	0.34	0.25	0.30	0.04	0.024	<0.005	-	-	-
41Cr4	0.36–0.45	0.80–1.20	0.17–0.37	max 0.30	0.50–0.90	max 0.05	max 0.035	max 0.035	max 0.05	max 0.10	max 0.20

2.2. Coating Preparation

To prepare the modifying solution, 4.84 mL of 3.16 mol dm^{-3} vinyltrimethoxysilane (VTMS) was used along with 2.16 mL of EtOH. The solution was placed on a magnetic stirrer and stirred for 2 days at 800 to 1000 rpm. A powder consisting of EDOT and PMo_{12} was then added. The procedure for obtaining PEDOT/PMo_{12} powder was described in a previous publication [60].

Different amounts of powder added to the silane solution were used in the experiment, i.e., 0.1 g, 0.15 g, 0.25 g, and 0.35 g. The solution was placed again on a magnetic stirrer for 24 h to obtain a gel consistency. The experiments were conducted at room temperature of 24 °C in laboratory settings.

VTMS/PEDOT/PMo_{12} coatings were applied to X20Cr13 and 41Cr4 steels by immersion for 20 min. The specimens were placed in a silica gel desiccator until the coatings dried completely (about 2 days).

2.3. Characteristics of Coatings

Microstructural examinations of coatings deposited on X20Cr13 and 41Cr4 steel were performed using a KEYENCE VHX—7000 digital microscope (Keyence, Mechelen, Belgium). The adhesion of VTMS/PEDOT/PMo_{12} coatings was tested using Scotch™ tape (Scotch™ Brand, St. Paul, MN, USA) with the pull-off method according to the ASTM D3359 standard [82].

The thickness of the coatings was measured using a Testan DT-20 AN 120 157 m (Anticorr, Gdańsk, Poland) with an integrated probe designed for ferromagnetic and non-ferromagnetic measurements. Corrosion resistance tests were carried out using a CH Instruments 706 measuring station (Austin, TX, USA). A three-electrode system was used: specimens of X20Cr13 and 41Cr4 steel without and with a coating: working electrode, counter-electrode (platinum wire), and reference electrode (saturated calomel electrode SCE). The properties of the anti-corrosion coatings were evaluated in two corrosion environments, 0.5 mol dm^{-3} Na_2SO_4 (pH = 2) and 0.5 mol dm^{-3} Na_2SO_4 + 0.5 mol dm^{-3} NaCl (pH = 2), using the Tafel Plot potentiodynamic technique, using potential scanning from −0.8 V to +1.6 V with a polarization rate of 10 mVs^{-1}. The tests were conducted at room temperature.

3. Results and Discussion

Figures 1 and 2 show the morphology of the VTMS/PEDOT/PMo_{12} coatings deposited on the surface of X20Cr13 and 41Cr4 steels. The surface and morphology of VTMS/PEDOT/PMo_{12} coatings deposited on X20Cr13 and 41Cr4 steels, at different concentrations of EDOT/PMo_{12} powder, were strongly dependent on the content of the components used in the deposition process. At the lowest powder concentration, 0.1 g (1a, 2a) and 0.15 g (1b, 2b) of EDOT/PMo_{12}, the VTMS/PEDOT/PMo_{12} coating on X20Cr13 steel showed a smooth and homogeneous surface. A low powder concentration leads to a thinner coating, which may exhibit poorer corrosion properties.

Increasing the EDOT/PMo_{12} powder content led to a more developed surface structure. As the concentration of EDOT/PMo_{12} increased, as with 0.25 g or 0.35 g, the coating became thicker and more opaque. The coatings exhibited a slightly rough structure, which increased the active surface area. The increase in powder concentration led to an increase in the thickness of the coating, which improved its ability to protect against external agents. The coating at 0.25 g or 0.35 g EDOT/PMo_{12} powder was characterized by a more complex morphology, with visible aggregates and pronounced surface irregularities. At this concentration, the coating can be expected to have better conductive properties due to the higher amount of polymer and heteropolyacid. However, higher roughness can also increase the risk of penetration by corrosive agents, which can limit the durability of the coating if adequate barrier properties are not provided.

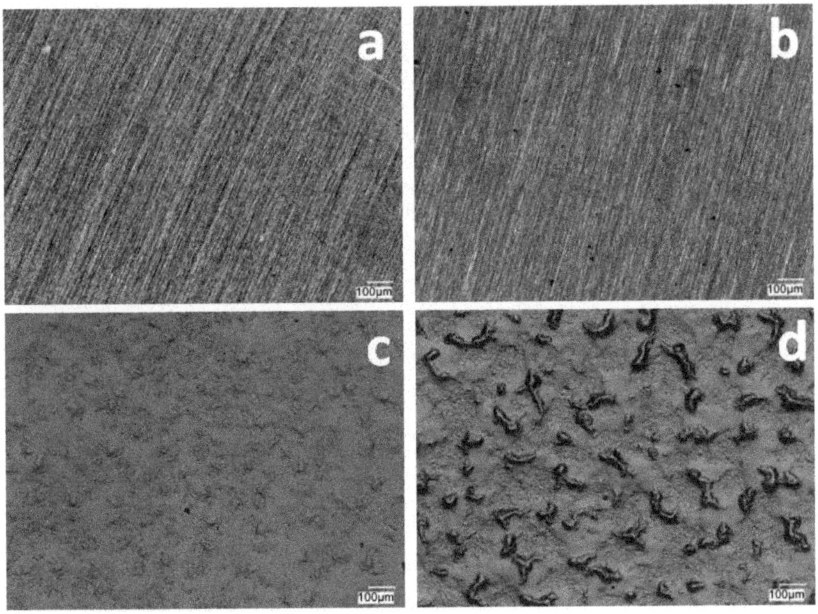

Figure 1. Surface of VTMS/PEDOT/PMo_{12} coatings deposited on X20Cr13 steel with VTMS concentration of 3.16 mol dm^{-3} and EDOT/PMo_{12} powder content: 0.1 g (**a**), 0.15 g (**b**), 0.25 g (**c**), and 0.35 g (**d**). KEYENCE VHX-7000 digital microscope, Magn. 200.

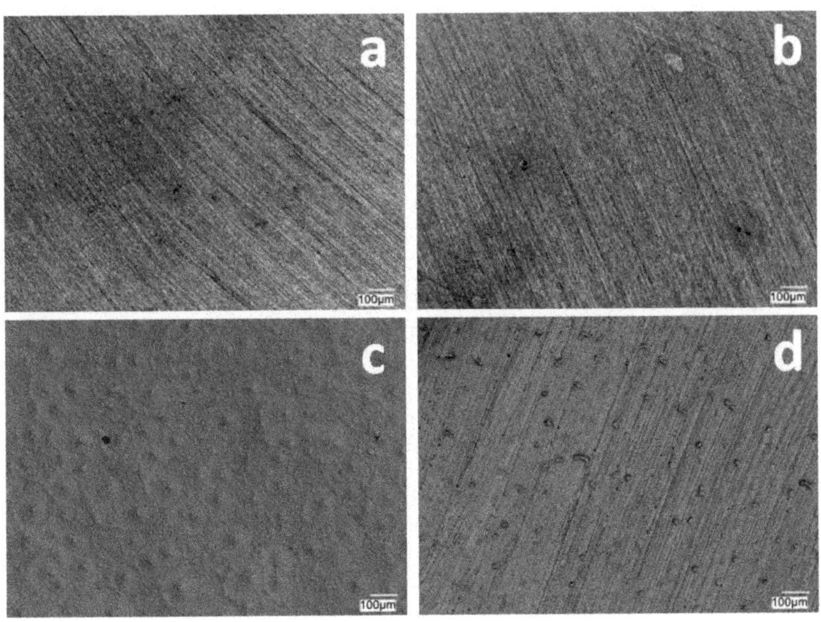

Figure 2. The surface of VTMS/PEDOT/PMo_{12} coatings deposited on 41Cr4 steel with VTMS concentration of 3.16 mol dm^{-3} and EDOT/PMo_{12} powder content: 0.1 g (**a**), 0.15 g (**b**), 0.25 g (**c**), and 0.35 g (**d**). KEYENCE VHX-7000 digital microscope, Magn. 200.

3.1. Scotch Test

To confirm the results of the adhesion of the VTMS/PEDOT/PMo_{12} coating, an adhesion test was performed using Scotch™ tape. The test was performed immediately after the deposition of VTMS/PEDOT/PMo_{12} coatings on the X20Cr13 and 41Cr4 steels. The tests conducted by the relevant standard showed that coatings containing 0.1 g and 0.15 g of EDOT/PMo12 powder demonstrated poor adhesion to the substrate. A material loss of approximately 30–40% of the coating was observed on X20Cr13 and 41Cr4 steels (classification 1B). In contrast, coatings with 0.25 g and 0.35 g of EDOT/PMo12 powder exhibited a material loss of less than 5% (classification 4B), indicating very good adhesion of the coatings to the substrate of both steels.

3.2. Thickness of the Coatings

One of the key parameters affecting the corrosion resistance of steel components is the thickness of protective coatings. In this study, this parameter was analyzed using two methods. Profiles were examined (Figures 3 and 4) to evaluate the thickness of the coatings. The average thickness of VTMS/PEDOT/PMo_{12} coatings with variable EDOT/PMo_{12} powder content (measured at 2 locations on the specimen) is presented in Table 2. The recorded thickness was measured using a KEYENCE digital microscope for different EDOT/PMo_{12} powder content: 0.1 g (a), 0.15 g (b), 0.25 g (c), and 0.35 g (d).

Table 2. Results of measurements of coating thickness using two instruments.

Instruments	Mean Thickness of VTMS/PEDOT/PMo_{12} Coating with Different EDOT/PMo_{12} Powder Content							
	0.1 g		0.15 g		0.25 g		0.35 g	
	Steel X20Cr13	Steel 41Cr4	Steel X20Cr13	Steel 41Cr4	Steel X20Cr13	Steel 41Cr4	Steel X20Cr13	Steel 41Cr4
KEYENCE VHX-7000 microscope	17.26 µm	7.24 µm	31.22 µm	13.97 µm	34.39 µm	22.14 µm	39.54 µm	35.86 µm
Testan DT-20 AN 120 157	16.58 µm	5.11 µm	29.59 µm	10.74 µm	32.81 µm	20.35 µm	37.27 µm	32.71 µm
Mean thickness (digital microscope and thickness gauge)	16.92 µm	6.16 µm	30.41 µm	12.36 µm	33.6 µm	21.25 µm	38.39 µm	34.29 µm

To compare coating thickness, in addition to the method described above, thickness measurements were performed using a Testan meter. A series of 10 measurements (at different sites on the specimens) was performed. Table 2 shows the average thickness for the individual VTMS/PEDOT/PMo_{12} coatings. The results are consistent with the thickness evaluated using a digital microscope. Based on measurements using two instruments (digital microscope and thickness gauge), the mean coating thickness was determined (Table 2).

The differences in coating thickness between the methods were small, confirming the suitability of the methods used to measure coating thickness. Based on the thickness measurements, it can be concluded that as the amount of EDOT/PMo_{12} powder in the coating increased, the thickness of the coating also increased.

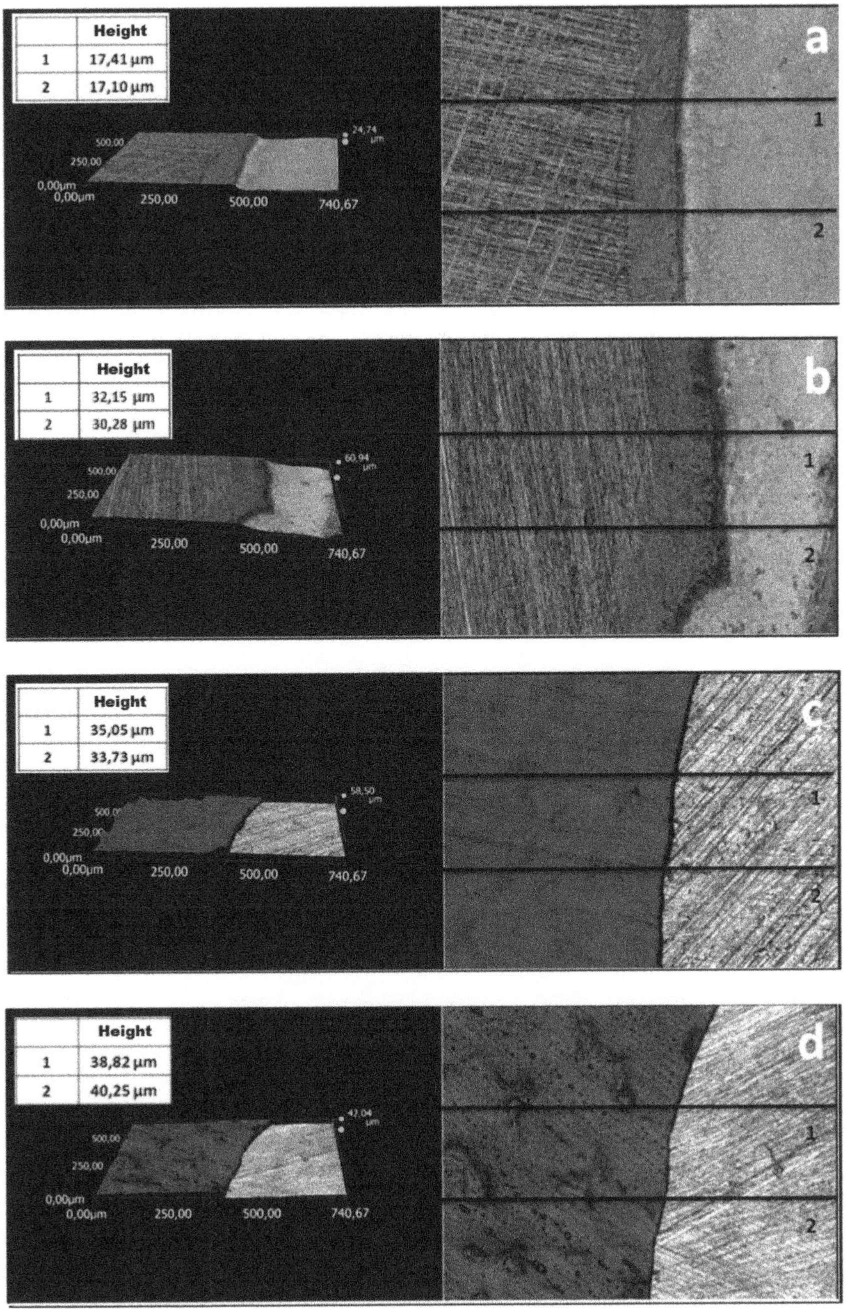

Figure 3. Surface profile of VTMS/PEDOT/PMo_{12} coatings deposited on X20Cr13 steel with different EDOT/PMo_{12} powder content: 0.1 g (**a**), 0.15 g (**b**), 0.25 g (**c**), 0.35 g (**d**).

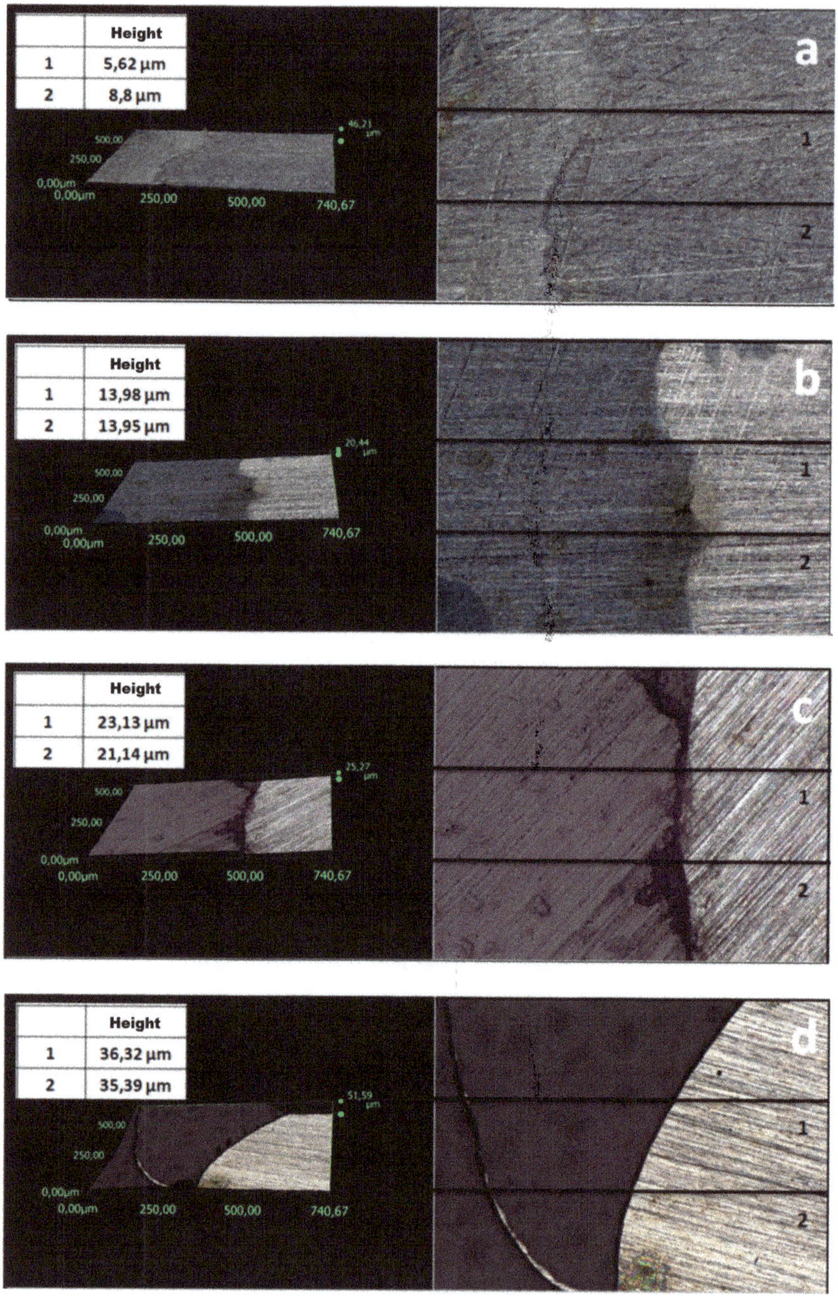

Figure 4. Surface profile of VTMS/PEDOT/PMo_{12} coatings deposited on 41Cr4 steel with different EDOT/PMo_{12} powder content: 0.1 g (**a**), 0.15 g (**b**), 0.25 g (**c**), 0.35 g (**d**).

3.3. Analysis of 2D Surface Geometry: Roughness Parameter

The analysis of 2D surface geometry was carried out by measuring the Ra parameter (Table 3). Test results varied depending on the content of the EDOT/PMo_{12} powder used in the modifying solution. As the content of the EDOT/PMo_{12} powder in the modifying solution increased, the value of the Ra parameter for the tested VTMS/PEDOT/PMo_{12} coatings also increased. The more EDOT/PMo_{12} powder in the modifying solution, the rougher the coating structure.

Table 3. Two-dimensional surface geometry analysis of VTMS/PEDOT/PMo_{12} coatings deposited on X20Cr13 and 41Cr4 steels: Ra parameter.

	2D Surface Roughness: Ra [μm]	
VTMS/PEDOT/PMo_{12} coating	X20Cr13 steel	41Cr4 steel
0.1 g of EDOT/PMo_{12} powder	3.66	1.83
0.15 g of EDOT/PMo_{12} powder	3.9	5.62
0.25 g of EDOT/PMo_{12} powder	5.24	6.97
0.35 g of EDOT/PMo_{12} powder	6.88	8.5

The closest Ra values were observed for VTMS/PEDOT/PMo_{12} coatings with EDOT/PMo_{12} powder contents of 0.1 g and 0.15 g and those with 0.25 g and 0.35 g for X20Cr13 steel. For 41Cr4 steel, the VTMS/PEDOT/PMo_{12} coatings with EDOT/PMo_{12} powder contents of 0.15 g and 0.25 g had the most similar Ra values. The lowest surface roughness was recorded for the VTMS/PEDOT/PMo_{12} coating with an EDOT/PMo_{12} powder content of 0.1 g for 41Cr4 steel.

3.4. Analysis of Coating Composition

The characterization of coatings based on vinyltrimethoxysilane VTMS, poly(3,4-ethylenedioxythiophene) PEDOT, and phosphomolybdic acid $H_3[Mo_{12}PO_{40}]*12H_2O$ deposited on steels (X20Cr13 and 41Cr4) was determined using the Fourier transform infrared spectrophotometer with attenuated total reflectance (QATR-FTIR) spectrophotometer method (Bruker, Bremen, Germany). A detailed description with drawings of the absorbance peaks for VTMS, PEDOT, and PMo_{12} coatings was presented in previous publications [47,50,57].

The VTMS/PEDOT/PMo_{12} coating spectrum cannot be considered a simple combination of VTMS, PEDOT, and PMo_{12} spectra. The characteristic bands of VTMS/PEDOT/PMo_{12} coatings have higher absorbance and wavelength shifts compared to VTMS, PEDOT, and PMo_{12} spectra. Due to the similar appearance (analogous course) of the other coatings, the FTIR spectrum for the VTMS/PEDOT/PMo_{12} coating with an EDOT/PMo_{12} powder content of 0.25 g (X20Cr13 steel) is shown in Figure 5.

Characteristic absorbance peaks observed for VTMS/PEDOT/PMo_{12} coatings in the 4000–400 cm^{-1} range are as follows:

- Absorbance bands observed at values of 3061 cm^{-1}, 2959 cm^{-1}, and 1275 cm^{-1} correspond to the asymmetric stretching and bending vibrations of the C-H bond belonging to the -Si-(OCH_3) group;
- Further peaks were recorded at 1604 cm^{-1} and 1408 cm^{-1}, which correspond to stretching vibrations of the C=C bond of the CH_2=CH-group;
- 1002 cm^{-1}, 885 cm^{-1}, and 742 cm^{-1} bands correspond to Si-O-C vibration;
- A broad band observed at 1190–1000 cm^{-1} corresponds to asymmetric stretching vibrations of Si-O-Si bonds;
- A peak at 692 cm^{-1} corresponds to the Si-C bond;
- A peak at 927 cm^{-1} was attributed to asymmetric bending vibrations of the Si-OH bond;

- 1021 cm^{-1}, 953 cm^{-1}, 852 cm^{-1}, and 805 cm^{-1}—four characteristic bands of the PMo_{12} ion, which can be related to asymmetric stretching with edge oxygen, respectively: P-O, Mo=O, Mo-O_c-Mo, and Mo-O_e-Mo;
- 583 cm^{-1}: the band attributed to the Keggin δ (O-P-O) structure;
- 529 cm^{-1} and 414 cm^{-1}: sharp Si-O bond bands.

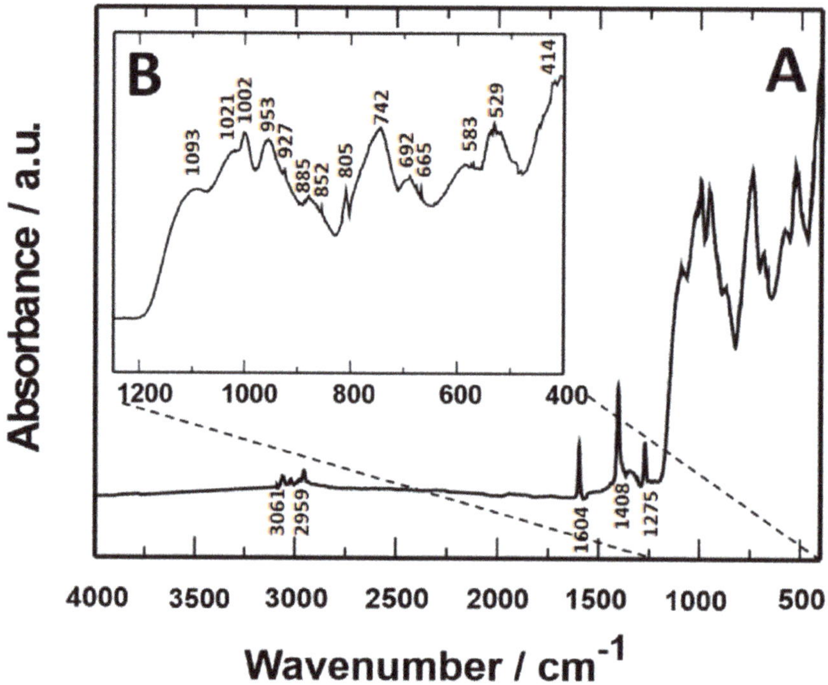

Figure 5. FTIR spectrum of VTMS/PEDOT/PMo_{12} coating with 0.25 g EDOT/PMo_{12} powder content, X20Cr13 steel substrate. (**A**) FTIR spectrum in the range 4000–400 cm^{-1}, (**B**) FTIR spectrum in the range 1200–400 cm^{-1}.

3.5. Corrosion Testing

Figures 6 and 7 show the open circuit potential (OCP) measurements recorded in solution: 0.5 mol dm^{-3} Na_2SO_4 pH = 2 and 0.5 mol dm^{-3} Na_2SO_4 + 0.5 mol dm^{-3} NaCl pH = 2 for X20Cr13 (Figure 6) and 41Cr4 (Figure 7) steels without (a) and with VTMS/PEDOT/PMo_{12} coating with EDOT/PMo_{12} powder content: 0.1 (b), 0.15 (c), 0.25 (d), and 0.35 (e) g. Uncoated X20Cr13 steel immediately after immersion in corrosive solutions showed potential of approximately −0.4 V. At longer exposure times, the corrosion potential value for the steel was −0.5 V. In contrast, 41Cr4 steel without coating immediately after immersion in corrosive solutions showed potential of −0.3 V to −0.6 V. For X20Cr13 and 41Cr4 steel with VTMS/PEDOT/PMo_{12} coatings, the potential remained in the range of 0.9 to −0.6 V, which was due to the different EDOT/PMo_{12} powder content (Table 4) in the sol–gel solution.

Figure 8 shows microscopic observations made after measuring OCP in a sulfate solution with the addition of Cl^- ions. VTMS/PEDOT/PMo_{12} coatings with EDOT/PMo_{12} powder contents of 0.1 g and 0.15 g showed local corrosion effects (pitting), indicating ineffective corrosion protection. No pitting corrosion was observed for the coating with the higher powder content in the sol–gel.

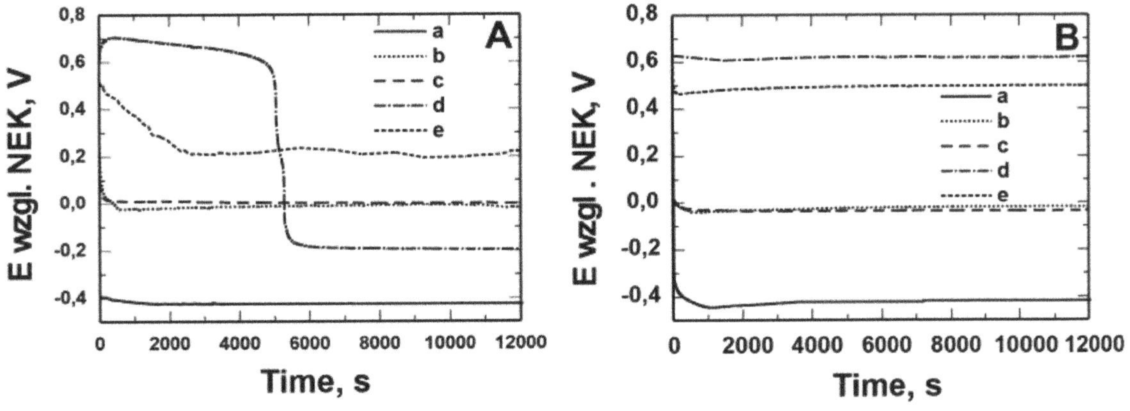

Figure 6. Measurement of the open potential of OCP from the time of exposure in solution: 0.5 mol dm^{-3} Na_2SO_4 pH = 2 (**A**) and 0.5 mol dm^{-3} Na_2SO_4 + 0.5 mol dm^{-3} NaCl pH = 2 (**B**) for X20Cr13 steel without (**a**) and with VTMS/PEDOT/PMo_{12} coatings with EDOT/PMo_{12} powder content: 0.1 g (**b**), 0.15 g (**c**), 0.25 g (**d**), 0.35 g (**e**).

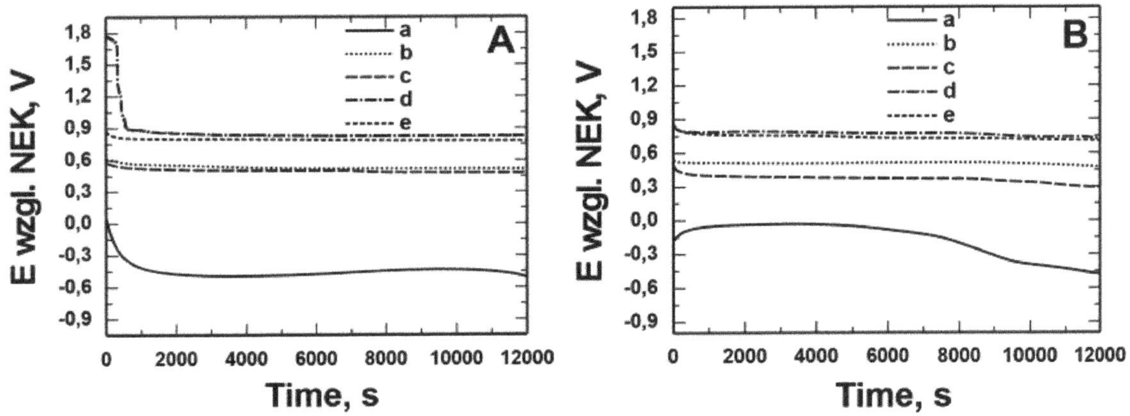

Figure 7. Measurement of the open potential of OCP from the time of exposure in solution: 0.5 mol dm^{-3} Na_2SO_4 pH = 2 (**A**) and 0.5 mol dm^{-3} Na_2SO_4 + 0.5 mol dm^{-3} NaCl pH = 2 (**B**) for 41Cr4 steel without (**a**) and with VTMS/PEDOT/PMo_{12} coatings with EDOT/PMo_{12} powder content: 0.1 g (**b**), 0.15 g (**c**), 0.25 g (**d**), 0.35 g (**e**).

To determine the most effective effect of EDOT/PMo_{12} powder content on the corrosion properties of the VTMS/PEDOT/PMo_{12} coatings on X20Cr13 and 41Cr4 steels, their ability to inhibit general and pitting corrosion was evaluated using potentiodynamic curves. Potentiodynamic measurements were performed in the log(i) = f(E) configuration and the curves were recorded at a polarization rate of 10 mVs^{-1}. This scan rate was sufficient to register Faradaic processes on the electrode. The total duration of a single measurement was very short, lasting 4 min.

Table 4. Results of measurements of open circuit potential (OCP) for different times of exposure in corrosive solutions.

Corrosion Solution	X20Cr13 Steel		41Cr4 Steel	
	EDOT/PMo_{12} [g]	Potential [V]	EDOT/PMo_{12} [g]	Potential [V]
general corrosion 0.5 mol dm^{-3} Na_2SO_4 ph = 2	0.1	ca. −0.15 V	0.1	ca. 0.6 V
	0.15	ca. 0.0 V	0.15	ca. 0.57 V
	0.25	initially at 0.7 V, rapidly decreasing to −0.2 V after 5500 s	0.25	falls sharply to 0.9 V for the first 1000 s and remains at this level for a further 11,000 s
	0.35	for about 2500 s decreases from 0.5 V to 0.2 V and remains at 0.2 V for another 9500 s	0.35	ca. 0.85 V
pitting corrosion 0.5 mol dm^{-3} Na_2SO_4 + 0.5 mol dm^{-3} NaCl pH = 2	0.1	ca. −0.1 V	0.1	0.55 V
	0.15	ca. −0.08 V	0.15	0.5 V to 0.3 V
	0.25	ca. 0.6 V	0.25	ca. 0.87 V
	0.35	ca. 0.5 V	0.35	ca 0.87 V

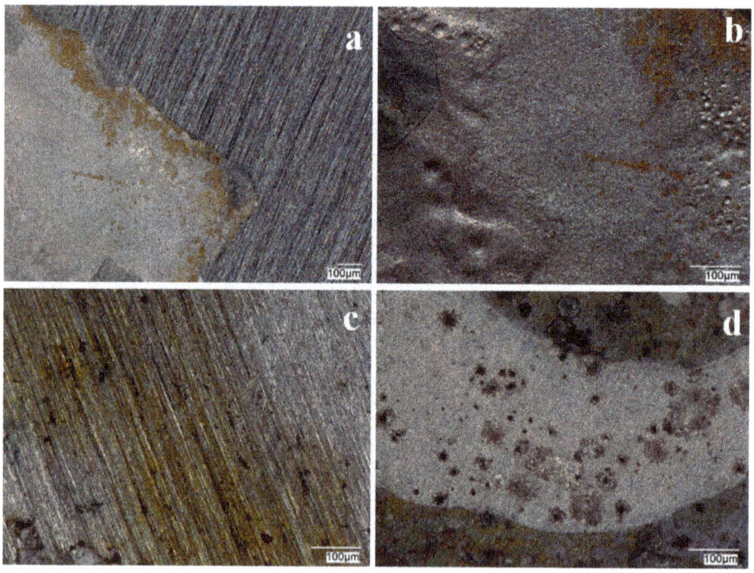

Figure 8. Photographs of VTMS/PEDOT/PMo_{12} coatings after OCP testing in a solution of 0.5 mol dm^{-3} Na_2SO_4 + 0.5 mol dm^{-3} NaCl pH = 2. Content of EDOT/PMo_{12} powder: 0.1 (**a**,**c**), 0.15 (**b**,**d**), substrate: X20Cr13 steel (**a**,**b**), 41Cr4 steel (**c**,**d**). KEYENCE VHX-7000 digital microscope, Magn. 200.

The experiment was conducted in two corrosion solutions: 0.5 mol dm^{-3} Na_2SO_4 pH = 2—general corrosion (Figures 9A and 10A) and 0.5 mol dm^{-3} Na_2SO_4 + 0.5 mol dm^{-3} NaCl pH = 2—pitting corrosion (Figures 9B and 10B). The potential range was $-0.8 \div 1.6$ V for X20Cr13 and 41Cr4 steel, uncoated and coated, respectively. As shown in Figures 9A and 10A, the VTMS/PEDOT/PMo_{12} coatings obtained in the study inhibited cathodic and anodic processes. There was a shift in the corrosion potential of the following aspects:

- X20Cr13 by approximately 0.9 V compared to uncoated steel (VTMS/PEDOT/PMo_{12} coating with 0.25 g of EDOT/PMo_{12} powder);

- 41Cr4 by approximately 0.7 V compared to uncoated steel (VTMS/PEDOT/PMo_{12} coating with 0.25 g of EDOT/PMo_{12} powder).

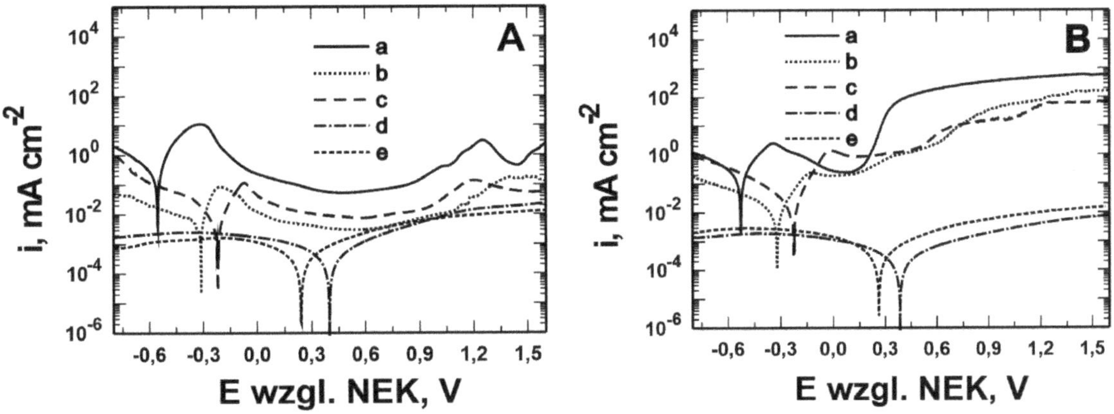

Figure 9. Potentiodynamic polarization curves recorded in the solution: 0.5 mol dm^{-3} Na_2SO_4 pH = 2 (**A**) and 0.5 mol dm^{-3} Na_2SO_4 + 0.5 mol dm^{-3} NaCl pH = 2 (**B**) for X20Cr13 steel without (**a**) and with VTMS/PEDOT/PMo_{12} coatings with powder content EDOT/PMo_{12} in solution: 0.1 (**b**), 0.15 (**c**), 0.25 (**d**), and 0.35 (**e**) g. Polarization rate 10 mVs^{-1}, solutions in contact with air.

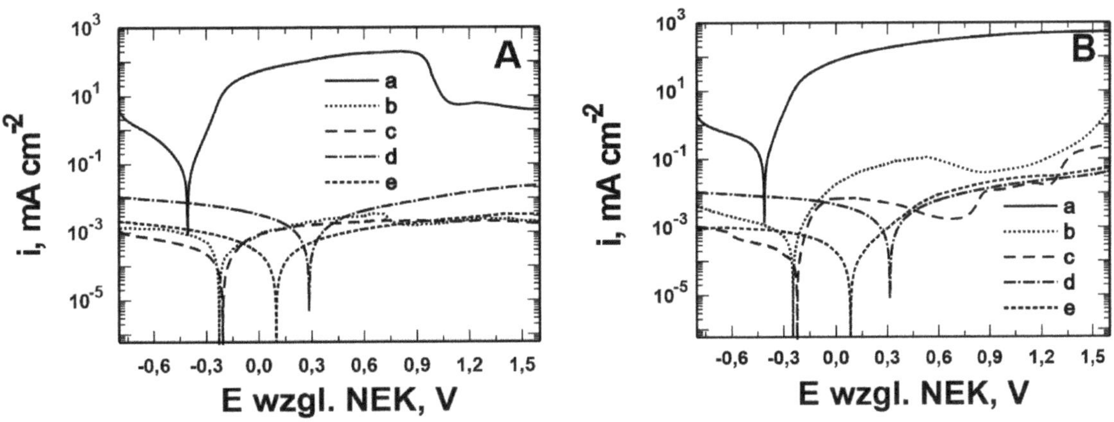

Figure 10. Potentiodynamic polarization curves recorded in the solution: 0.5 mol dm^{-3} Na_2SO_4 pH = 2 (**A**) and 0.5 mol dm^{-3} Na_2SO_4 + 0.5 mol dm^{-3} NaCl pH = 2 (**B**) for 41Cr4 steel without (**a**) and with VTMS/PEDOT/PMo_{12} coatings with powder content EDOT/PMo_{12} in solution: 0.1 (**b**), 0.15 (**c**), 0.25 (**d**), and 0.35 (**e**) g. Polarization rate 10 mVs^{-1}, solutions in contact with air.

Anodic current densities for both steels with VTMS/PEDOT/PMo_{12} coatings in the passive range were 1 to 3 times lower than for uncoated steel.

Table 5 presents the data obtained from the Tafel plot and the polarization resistance calculated for X20Cr13 steel.

For the 41Cr4 steel, the calculated polarization resistance values are as follows—for general corrosion, the following values were obtained for coating powders: 0.1 g–15.5007 Ωcm^2, 0.15 g–26.3226 Ωcm^2, 0.25 g–39.4664 Ωcm^2, and 0.35 g–84.7618 Ωcm^2; and for pitting corrosion: 0.1 g–9.7513 Ωcm^2, 0.15 g–15.731 Ωcm^2, 0.25 g–27.1054 Ωcm^2, and 0.35 g–99.615 Ωcm^2. The data obtained show that the polarization resistance was highest for both steels coated with 0.35 g of powder, indicating good corrosion protection.

Table 5. Data obtained from the Tafel plot and polarization resistance calculated for X20Cr13 steel.

Steel X20Cr13 VTMS/PEDOT/PMo$_{12}$	Potential Corrosive E [V]		Current Density Corrosive i [mAcm^{-2}]		Polarization Resistance tgα [Ω*cm^2]	
	General Corrosion	Pitting Corrosion	General Corrosion	Pitting Corrosion	General Corrosion	Pitting Corrosion
0.1 g	−0.313	−0.319	8.08×10^{-5}	9.765×10^{-5}	0.50168	0.66663
0.15 g	−0.220	−0.219	6.705×10^{-5}	4.8155×10^{-4}	1.05892	0.26357
0.25 g	0.398	0.415	9.115×10^{-8}	9.725×10^{-5}	36.55406	61.10981
0.35 g	0.242	0.244	6.78×10^{-6}	9.825×10^{-5}	43.72182	85.44327

To evaluate the ability of the obtained coatings to inhibit pitting corrosion, potentiodynamic curves were obtained for a solution of 0.5 mol dm^{-3} Na_2SO_4 + 0.5 mol dm^{-3} NaCl pH = 2 (Figures 9B and 10B). The corrosion potential of X20Cr13 steel for all coatings shifted by about 0.2–0.9 V toward positive values compared to the corrosion potential recorded for uncoated steel (E_{corr} = −0.527 V). For 41Cr4 steel, the corrosion potential shifted toward positive values by approximately 0.2–0.7 V compared to the corrosion potential recorded for uncoated steel (E_{corr} = −0.45 V). Lower cathodic and anodic current density values were observed for both steels compared to uncoated steels. The polarization curves show that the pitting nucleation potential (E_{pit}) was 0.12 V for uncoated X20Cr13 steel, whereas for the steel coated with VTMS/PEDOT/PMo$_{12}$ with EDOT/PMo$_{12}$ powder content, it was: 0.3 V for 0.1 g and 0.75 V for 0.15 g. For 41Cr4 steel without coating, no pitting nucleation potential (E_{pit}) was observed, while for VTMS/PEDOT/PMo$_{12}$ coating, it was 0.6 V for 0.1 g and 0.9 V for 0.15 g. For the VTMS/PEDOT/PMo$_{12}$ coating with 0.25 g and 0.35 g of EDOT/PMo$_{12}$ powder, no pitting nucleation potential was observed.

The silane coating with a conductive polymer and Keggin-type acid (heteropolyacid) effectively inhibited the access of aggressive anions to the steel substrate, thereby protecting the substrate from pitting corrosion. Microscopic observations made after the measurement showed no local corrosion effects under the VTMS/PEDOT/PMo$_{12}$ coating for EDOT/PMo$_{12}$ powder contents of 0.25 and 0.35 mol dm^{-3} for both steels (Figures 11 and 12). The coating with 0.1 and 0.15 g of EDOT/PMo$_{12}$ powder showed local pitting corrosion sites (Figures 10 and 11a,b). The high content of EDOT/PMo$_{12}$ powder in the coating affected the higher density of the modification solution, resulting in a thicker coating and increased roughness in the coating structure.

A chronoamperometric method was used to verify the resistance to pitting corrosion of coatings deposited on steels (Figure 13A,B). In this method, changes in current density are recorded versus time when a constant potential is applied to the working electrode. Pitting nucleation can be inferred from the chronoamperometric curves. To determine the stability of the applied coatings, the time of the specimens remaining in the corrosive solution containing chloride ions and current density at the set potential were compared. Chronoamperometric curves were recorded in a solution of 0.5 mol dm^{-3} Na_2SO_4 + 0.5 mol dm^{-3} NaCl pH = 2 at a potential of 0.1 V for X20Cr13 steel and 0.5 V for 41Cr4 steel. As can be observed, the initiation of pitting formation on both steels occurred within a few seconds, followed by a rapid increase in current density. For VTMS/PEDOT/PMo$_{12}$ coatings with EDOT/PMo$_{12}$ powder contents of 0.25 and 0.35 g, the highest corrosion resistance was obtained for both steels. No increase in current density was observed for the above coatings during the 200 h measurement. An increase in current density was recorded over the time interval as follows:

- 50 to 100 h for VTMS/PEDOT/PMo$_{12}$ coating with 0.1 g of EDOT/PMo$_{12}$ powder for X20Cr13 and 41Cr4 steels;
- 120 to 125 h for VTMS/PEDOT/PMo$_{12}$ coating with 0.15 g of EDOT/PMo$_{12}$ powder for X20Cr13 steel;
- 125 to 160 h for VTMS/PEDOT/PMo$_{12}$ coating with 0.15 g of EDOT/PMo$_{12}$ powder for 41Cr4 steel.

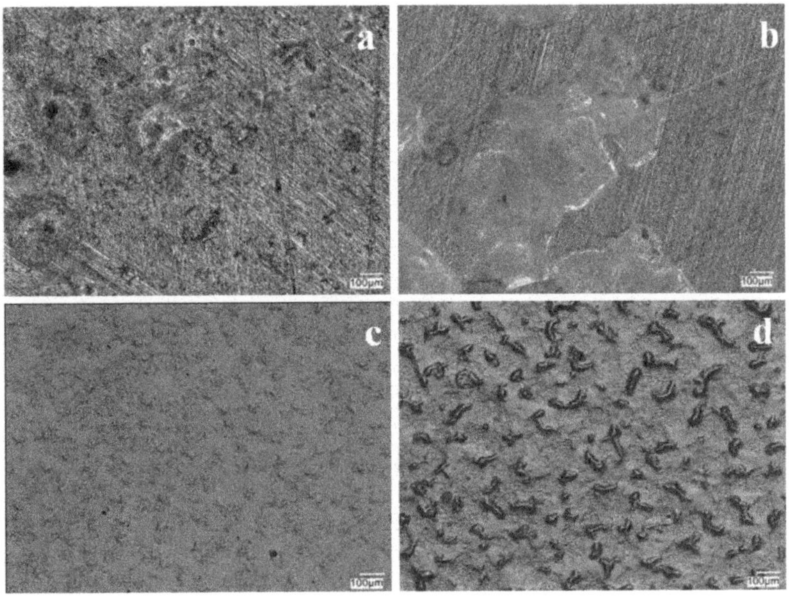

Figure 11. Photographs of VTMS/PEDOT/PMo_{12} coatings after potentiodynamic testing in a corrosion solution of 0.5 mol dm^{-3} Na_2SO_4 + 0.5 mol dm^{-3} NaCl pH = 2. Content of EDOT/PMo_{12} powder: 0.1 (**a**), 0.15 (**b**), 0.25 (**c**), and 0.35 (**d**) g, substrate X20Cr13 steel. KEYENCE VHX-7000 digital microscope, Magn. 200.

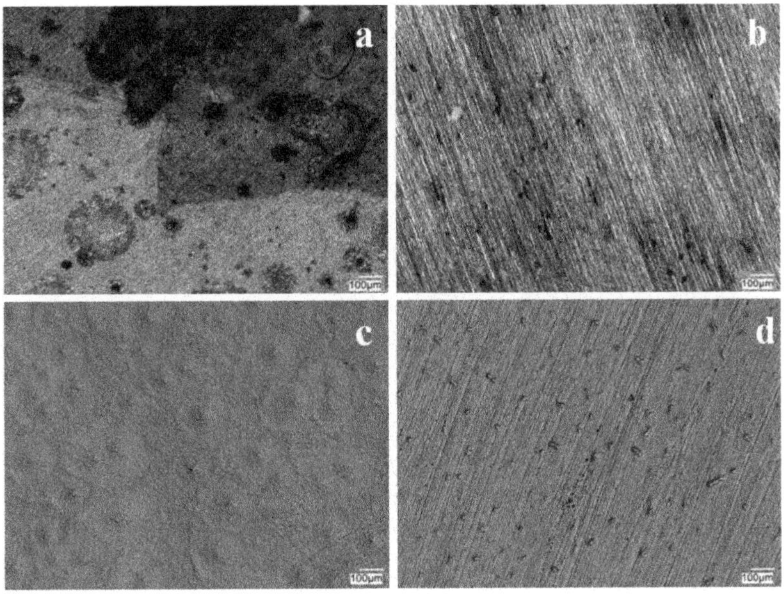

Figure 12. Photographs of VTMS/PEDOT/PMo_{12} coatings after potentiodynamic testing in a corrosion solution of 0.5 mol dm^{-3} Na_2SO_4 + 0.5 mol dm^{-3} NaCl pH = 2. Content of EDOT/PMo_{12} powder: 0.1 (**a**), 0.15 (**b**), 0.25 (**c**), and 0.35 (**d**) g, substrate 41Cr4 steel. KEYENCE VHX-7000 digital microscope, Magn. 200.

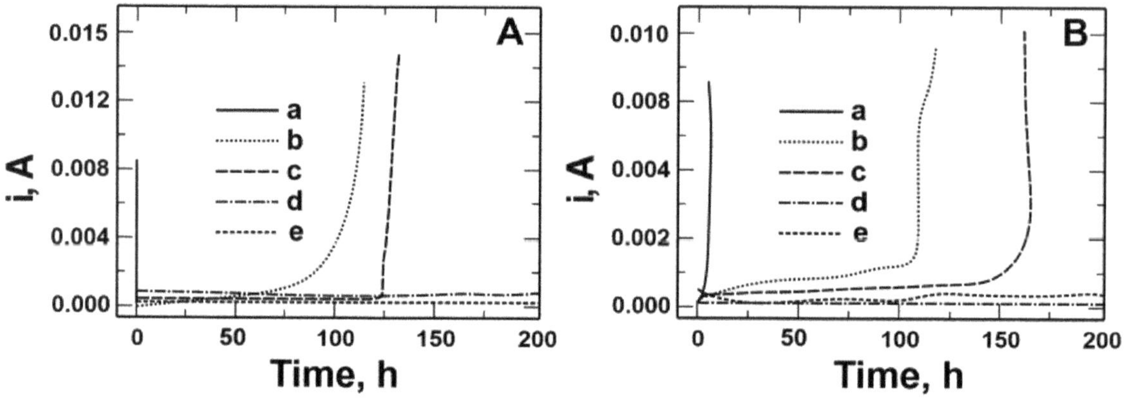

Figure 13. Chronoamperometric curves recorded in a solution of 0.5 mol dm^{-3} Na_2SO_4 + 0.5 mol dm^{-3} NaCl pH = 2 for steel: X20Cr13 (**A**) and 41Cr4 (**B**) without (**a**) and with VTMS/PEDOT/PMo_{12} coatings with EDOT/PMo_{12} powder content: 0.1 (**b**), 0.15 (**c**), 0.25 (**d**), and 0.35 (**e**) g.

The best ability to block the transport of chloride ions responsible for pitting corrosion was demonstrated by VTMS/PEDOT/PMo_{12} coatings with 0.25 and 0.35 g of EDOT/PMo_{12} powder (200 h).

4. Conclusions

In the present study, anticorrosion coatings based on vinyltrimethoxysilane (VTMS), poly(3,4-ethylenedioxythiophene) (PEDOT), and phosphododecamolybdic acid (PMo_{12}) obtained on X20Cr13 and 41Cr4 steel by the sol–gel method were shown to be highly effective in terms of corrosion protection. The use of the immersion method allowed for coatings with high adhesion to be obtained, especially for specimens containing 0.25 g and 0.35 g of EDOT/PMo_{12} powder, which showed the best protective properties. These coatings were homogeneous, without cracks or defects, and their thickness depended on the amount of modifier powder added. The results of the surface morphology tests showed a low degree of roughness, an important factor in favor of corrosion resistance, especially for the specimens with 0.1 g of EDOT/PMo_{12} in the VTMS/PEDOT/PMo_{12} coating, where the lowest roughness (Ra) was recorded.

FTIR analysis confirmed the presence of significant chemical bonds, such as C-H, C=C, Si-O-C, Si-O-Si, P-O, and Mo=O, which are indicative of proper crosslinking of the coatings. Corrosion resistance tests were carried out in sulfate solutions, both without and with Cl- ions acidified to pH 2. These results indicated that the coatings effectively protected the steel against general and pitting corrosion, stabilized the corrosion potential in the passive state, and provided anodic barrier protection. Particularly effective were coatings with 0.25 g and 0.35 g of EDOT/PMo_{12} powder, which showed no signs of pitting corrosion even in the presence of chloride ions. OCP and chronoamperometric test results further confirmed the high resistance of these coatings to various forms of corrosion.

Based on the test results, it can be concluded that the proposed VTMS/PEDOT/PMo_{12} coatings significantly increase the corrosion resistance of X20Cr13 and 41Cr4 steels. Their protective properties, homogeneity, and stability in aggressive environments indicate their potential use in various industries where high corrosion resistance is crucial.

The coating described in this article can be used as a protective barrier in painting systems designed to protect steel structures. Due to its chemical stability and insolubility, it can provide effective protection against harmful external factors such as moisture and aggressive chemicals that can lead to steel corrosion.

The proposed protective coating, as demonstrated by the studies, exhibits high corrosion resistance, which can significantly extend the service life of steel structures, minimizing the need for frequent maintenance and repairs.

Thanks to its protective function, this coating could also serve as a foundation for the development of new material technologies capable of ensuring even better durability and efficiency, including in more demanding industrial conditions, such as environmental protection systems or catalytic processes in the chemical industry.

As part of future work, we plan to conduct further research on the use of other heteropolyacid structures, particularly those with much higher charge, such as Dawson-type heteropolyacids, as well as cesium salts of vanadium-substituted polyoxometalates, which are more stable and insoluble. These studies may contribute to the development of more resistant and efficient materials with improved catalytic and structural properties.

Additionally, compared to our previous studies [83] on silane coatings, it has been shown that the addition of EDOT/PMo12 powder to silane significantly improves the corrosion protection of the substrate. Furthermore, the results presented in the work [63] also show that the coating composed of EDOT/PMo_{12} exhibits lower corrosion resistance compared to the coating proposed in the current article. These facts unequivocally confirm the correctness of the research direction chosen by the authors, especially regarding the modification of the coating structure to improve the corrosion resistance of the substrate.

Author Contributions: Conceptualization, L.A. and K.M.; methodology, L.A., K.M. and A.K.-K.; formal analysis, L.A., K.M. and A.D; investigation, A.K.-K., L.A. and K.M.; data curation, L.A. and K.M.; writing—original draft preparation, A.K.-K., L.A., K.M. and A.D.; writing—review and editing, A.K.-K., L.A., K.M. and A.D.; visualization, L.A., K.M. and A.D.; supervision, L.A., K.M. and A.D.; project administration, L.A. and K.M. All authors have read and agreed to the published version of the manuscript.

Funding: This research received no external funding.

Institutional Review Board Statement: Not applicable.

Informed Consent Statement: Not applicable.

Data Availability Statement: The original contributions presented in this study are included in the article. Further inquiries can be directed to the corresponding author.

Conflicts of Interest: The authors declare no conflicts of interest.

References

1. Cobb, H.M. *The History of Stainless Steel*; ASM International: Northeast Ohio, OH, USA, 2010.
2. Rossi, B. Discussion on the use of stainless steel in constructions in view of sustainability. *Thin-Walled Struct.* **2014**, *83*, 182–189. [CrossRef]
3. Gardner, L. Stability and design of stainless steel structures–Review and outlook. *Thin-Walled Struct.* **2019**, *141*, 208–216. [CrossRef]
4. Gardner, L. The use of stainless steel in structures. *Prog. Struct. Eng. Mater.* **2005**, *7*, 45–55. [CrossRef]
5. Dutta, S. Different types and new applications of stainless steel. *Stainl. Steel* **2018**, *62*, 86–91.
6. McGuire, M.F. *Stainless Steels for Design Engineers*; ASM International: Northeast Ohio, OH, USA, 2008.
7. Birtărescu, E.; Nedeloni, M.D.; Pedrali, P.C.; Câmpian, C.V.; Nedeloni, L.; Ene, T.; Bogdan, S.L. Some laboratory tests regarding the X20Cr13 martensitic stainless steel behaviour. In *IOP Conference Series: Materials Science and Engineering*; IOP Publishing: Bristol, UK, 2018; Volume 416, p. 012025.
8. Dewidar, M.M.; Khalil, K.A.; Lim, J.K. Processing and mechanical properties of porous 316L stainless steel for biomedical applications. *Trans. Nonferrous Met. Soc. China* **2007**, *17*, 468–473. [CrossRef]
9. Kostadin, T.; Jakovljevic, S. Analysis of corrosion resistance when turning martensitic stainless steel X20Cr13 under chilled air-cooling. *Adv. Prod. Eng. Manag.* **2017**, *12*, 105–114. [CrossRef]
10. Ryan, M.P.; Williams, D.E.; Chater, R.J.; Hutton, B.M.; McPhail, D.S. Why stainless steel corrodes. *Nature* **2002**, *415*, 770–774. [CrossRef]
11. Kadry, S. Corrosion analysis of stainless steel. *Eur. J. Sci. Res.* **2008**, *22*, 508–516.
12. Xavier, J.R. Investigation on the anticorrosion, adhesion and mechanical performance of epoxy nanocomposite coatings containing epoxy-silane treated nano-MoO_3 on mild steel. *J. Adhes. Sci. Technol.* **2020**, *34*, 115–134. [CrossRef]

13. Wang, L.; Liu, C.S.; Yu, H.Y.; An, C.Q. Structure and corrosion resistance of a composite γ-amino propyl triethoxy silane and γ-glycidoxy propyl trimethoxy silane conversion coating on galvanized steel. *J. Iron Steel Res. Int.* **2012**, *19*, 46–51. [CrossRef]
14. Fedel, M.; Olivier, M.; Poelman, M.; Deflorian, F.; Rossi, S.; Druart, M.E. Corrosion protection properties of silane pre-treated powder coated galvanized steel. *Prog. Org. Coat.* **2009**, *66*, 118–128. [CrossRef]
15. Pape, P.G. Adhesion promoters: Silane coupling agents. In *Applied Plastics Engineering Handbook*; William Andrew Publishing: Norwich, NY, USA, 2011; pp. 503–517.
16. Walker, P. Silane and other adhesion promoters in adhesive technology. *Handb. Adhes. Technol. N. Y. Taylor Fr. Group* **2003**, 205–222.
17. Matisons, J.G. Silane coupling agents and glass fibre surfaces: A perspective. *Silanes Other Coupling Agents* **2009**, *5*, 3–22.
18. Mittal, K.L. *Silanes and Other Coupling Agents*; CRC Press: Leiden, The Netherlands; Boston, MA, USA, 2007; Volume 4.
19. Lung, C.Y.K.; Matinlinna, J.P. Aspects of silane coupling agents and surface conditioning in dentistry: An overview. *Dent. Mater.* **2012**, *28*, 467–477. [CrossRef] [PubMed]
20. Lo, T.N.; Lee, J.; Hwang, H.S.; Park, I. Nanoscale coatings derived from fluoroalkyl and PDMS alkoxysilanes on rough aluminium surfaces for improved durability and anti-icing properties. *ACS Appl. Nano Mater.* **2021**, *4*, 7493–7501. [CrossRef]
21. Tshabalala, M.A.; Kingshott, P.; VanLandingham, M.R.; Plackett, D. Surface chemistry and moisture sorption properties of wood coated with multifunctional alkoxysilanes by sol-gel process. *J. Appl. Polym. Sci.* **2003**, *88*, 2828–2841. [CrossRef]
22. Romero-Gavilán, F.; Carlos-Almeida, J.; Cerqueira, A.; Gurruchaga, M.; Goni, I.; Miranda-Salvado, I.M.; Suay, J. Sol-gel coatings made using methyl-modified alkoxysilanes: The balance between protection and bioactivation. *Prog. Org. Coat.* **2020**, *147*, 105770. [CrossRef]
23. Wang, S.; Mahlberg, R.; Jämsä, S.; Nikkola, J.; Mannila, J.; Ritschkoff, A.C.; Peltonen, J. Surface properties and moisture behaviour of pine and heat-treated spruce modified with alkoxysilanes by sol–gel process. *Prog. Org. Coat.* **2011**, *71*, 274–282. [CrossRef]
24. Yaneff, P.V.; Adamsons, K.; Ryntz, R.A.; Britz, D. Structure/property relationships in flexible alkoxysilane automotive coatings. *J. Coat. Technol.* **2002**, *74*, 135–141. [CrossRef]
25. Sardon, H.; Irusta, L.; González, A.; Fernández-Berridi, M.J. Waterborne hybrid polyurethane coatings functionalized with (3-aminopropyl) triethoxysilane: Adhesion properties. *Prog. Org. Coat.* **2013**, *76*, 1230–1235. [CrossRef]
26. Jang, J.; Kim, E.K. Corrosion protection of epoxy-coated steel using different silane coupling agents. *J. Appl. Polym. Sci.* **1999**, *71*, 585–593. [CrossRef]
27. Plueddemann, E.P. Reminiscing on silane coupling agents. *J. Adhes. Sci. Technol.* **1991**, *5*, 261–277. [CrossRef]
28. Boerio, F.J.; Edwin, P. Plueddemann: Experimental laboratory chemist. *J. Adhes. Sci. Technol.* **1991**, *5*, 253–260. [CrossRef]
29. Blum, F.D.; Meesiri, W.; Kang, H.J.; Gambogi, J.E. Hydrolysis, adsorption, and dynamics of silane coupling agents on silica surfaces. *J. Adhes. Sci. Technol.* **1991**, *5*, 479–496. [CrossRef]
30. Ahmad, N.; MacDiarmid, A.G. Inhibition of corrosion of steels with the exploitation of conducting polymers. *Synth. Met.* **1996**, *78*, 103–110. [CrossRef]
31. Racicot, R.; Clark, R.L.; Liu, H.B.; Yang, S.C.; Alias, M.N.; Brown, R. Thin film conductive polymers on aluminum surfaces: Interfacial charge-transfer and anticorrosion aspects. In *Optical and Photonic Applications of Electroactive and Conducting Polymers*; SPIE's 1995 International Symposiumon Optical Science, Engineering and Instrumentation; SPIE: San Diego, CA, USA, 1995; Volume 2528, pp. 251–258.
32. Tallman, D.E.; Pae, Y.; Bierwagen, G.P. Conducting polymers and corrosion: Polyaniline on steel. *Corrosion* **1999**, *55*, 779–786. [CrossRef]
33. Zarras, P.; Anderson, N.; Webber, C.; Irvin, D.J.; Irvin, J.A.; Guenthner, A.; Stenger-Smith, J.D. Progress in using conductive polymers as corrosion-inhibiting coatings. *Radiat. Phys. Chem.* **2003**, *68*, 387–394. [CrossRef]
34. Zubair, N.A.; Rahman, N.A.; Lim, H.N.; Sulaiman, Y. Production of conductive PEDOT-coated PVA-GO composite nanofibers. *Nanoscale Res. Lett.* **2017**, *12*, 113. [CrossRef]
35. Ohtsuka, T. Corrosion protection of steels by conducting polymer coating. *Int. J. Corros.* **2012**, *1*, 915090. [CrossRef]
36. Adamczyk, A. Polimery przewodzące w ochronie przed korozją. *Ochr. Przed Korozją* **2018**, *12*, 363–368. [CrossRef]
37. Luo, S.C.; Yu, H.H.; Wan, A.C.; Han, Y.; Ying, J.Y. A General Synthesis for PEDOT-Coated Nonconductive Materials and PEDOT Hollow Particles by Aqueous Chemical Polymerization. *Small* **2008**, *4*, 2051–2058. [CrossRef] [PubMed]
38. Kumar, A.M.; Hussein, M.A.; Adesina, A.Y.; Ramakrishna, S.; Al-Aqeeli, N. Influence of surface treatment on PEDOT coatings: Surface and electrochemical corrosion aspects of newly developed Ti alloy. *RSC Adv.* **2018**, *8*, 19181–19195. [CrossRef] [PubMed]
39. Vreeland, R.F.; Atcherley, C.W.; Russell, W.S.; Xie, J.Y.; Lu, D.; Laude, N.D.; Heien, M.L. Biocompatible PEDOT: Nafion composite electrode coatings for selective detection of neurotransmitters in vivo. *Anal. Chem.* **2015**, *87*, 2600–2607. [CrossRef]
40. Rohwerder, M. Conducting polymers for corrosion protection: A review. *Int. J. Mater. Res.* **2009**, *100*, 1331–1342. [CrossRef]
41. Bazli, L.; Yusuf, M.; Farahani, A.; Kiamarzi, M.; Seyedhosseini, Z.; Nezhadmansari, M.; Iranpoor, M. Application of composite conducting polymers for improving the corrosion behavior of various substrates: A Review. *J. Compos. Compd.* **2020**, *2*, 228–240. [CrossRef]
42. Adamczyk, L.; Dudek, A. *The Influence of Components Concentration in the Electrodeposition Process on the Protective Properties of 3,4-ethylenedioxythiophene (EDOT) with 4-(pyrrole-1-yl) Benzoil Acid (PYBA), Polyoxyethylene-10-laurylethier (BRIJ) and Lithium Perchlorate*; EU: Brno, Czech Republic, 2019.
43. Zhu, D.; Lu, X.; Lu, Q. Electrically conductive PEDOT coating with self-healing superhydrophobicity. *Langmuir* **2014**, *30*, 4671–4677. [CrossRef]

44. Boehler, C.; Aqrawe, Z.; Asplund, M. Applications of PEDOT in bioelectronic medicine. *Bioelectron. Med.* **2019**, *2*, 89–99. [CrossRef]
45. Zhang, Z.; Tian, G.; Duan, X.; Chen, H.L.; Kim Richie, D.H. Nanostructured PEDOT Coatings for Electrode–Neuron Integration. *ACS Appl. Bio Mater.* **2021**, *4*, 5556–5565. [CrossRef]
46. Heydari Gharahcheshmeh, M.; Wan, C.T.C.; Ashraf Gandomi, Y.; Greco, K.V.; Forner-Cuenca, A.; Chiang, Y.M.; Gleason, K.K. Ultrathin conformal oCVD PEDOT coatings on carbon electrodes enable improved performance of redox flow batteries. *Adv. Mater. Interfaces* **2020**, *7*, 2000855. [CrossRef]
47. Gueye, M.N.; Carella, A.; Faure-Vincent, J.; Demadrille, R.; Simonato, J.P. Progress in understanding structure and transport properties of PEDOT-based materials: A critical review. *Prog. Mater. Sci.* **2020**, *108*, 100616. [CrossRef]
48. Namsheer, K.; Rout, C.S. Conducting polymers: A comprehensive review on recent advances in synthesis, properties and applications. *RSC Adv.* **2021**, *11*, 5659–5697.
49. Schultze, J.W.; Karabulut, H. Application potential of conducting polymers. *Electrochim. Acta* **2005**, *50*, 1739–1745. [CrossRef]
50. Wang, X.S.; Tang, H.P.; Li, X.D.; Hua, X. Investigations on the mechanical properties of conducting polymer coating-substrate structures and their influencing factors. *Int. J. Mol. Sci.* **2009**, *10*, 5257–5284. [CrossRef] [PubMed]
51. Riaz, U.; Nwaoha, C.; Ashraf, S.M.; Laboratory, M.R. Recent advances in corrosion protective composite coatings based on conducting polymers and natural resource derived polymers. *Prog. Org. Coat.* **2014**, *77*, 743–756. [CrossRef]
52. Guimard, N.K.; Gomez, N.; Schmidt, C.E. Conducting polymers in biomedical engineering. *Prog. Polym. Sci.* **2007**, *32*, 876–921. [CrossRef]
53. Kowalski, D.; Ueda, M.; Ohtsuka, T. Self-healing ion-permselective conducting polymer coating. *J. Mater. Chem.* **2010**, *20*, 7630–7633. [CrossRef]
54. McFarlane, S.L.; Deore, B.A.; Svenda, N.; Freund, M.S. A one-step, organic-solvent processable synthesis of PEDOT thin films via in situ metastable chemical polymerization. *Macromolecules* **2010**, *43*, 10241–10245. [CrossRef]
55. Ouyang, L.; Kuo, C.C.; Farrell, B.; Pathak, S.; Wei, B.; Qu, J.; Martin, D.C. Poly [3,4-ethylene dioxythiophene (EDOT)-co-1,3,5-tri [2-(3,4-ethylene dioxythienyl)]-benzene (EPh)] copolymers (PEDOT-co-EPh): Optical, electrochemical and mechanical properties. *J. Mater. Chem. B* **2015**, *3*, 5010–5020. [CrossRef]
56. Wei, B.; Liu, J.; Ouyang, L.; Kuo, C.C.; Martin, D.C. Significant enhancement of PEDOT thin film adhesion to inorganic solid substrates with EDOT-acid. *ACS Appl. Mater. Interfaces* **2015**, *7*, 15388–15394. [CrossRef]
57. Zuber, K.; Fabretto, M.; Hall, C.; Murphy, P. Improved PEDOT conductivity via suppression of crystallite formation in Fe (III) tosylate during vapor phase polymerization. *Macromol. Rapid Commun.* **2008**, *29*, 1503–1508. [CrossRef]
58. Kucharczyk, A.; Adamczyk, L. Wpływ dodatku polimeru przewodzącego (PEDOT, PANI) na właściwości ochronne powłok antykorozyjnych na bazie silanu (VTMS) i surfaktantu (BRIJ) osadzonych na stali $X_{20}Cr_{13}$. *Poszerzamy Horyzonty* **2022**, *32*, 203–210.
59. Karnicka, K.; Chojak, M.; Miecznikowski, K.; Skunik, M.; Baranowska, B.; Kolary, A.; Kulesza, P.J. Polyoxometallates as inorganic templates for electrocatalytic network films of ultra-thin conducting polymers and platinum nanoparticles. *Bioelectrochemistry* **2005**, *66*, 79–87. [CrossRef] [PubMed]
60. Adamczyk, L.; Kulesza, P.J.; Miecznikowski, K.; Palys, B.; Chojak, M.; Krawczyk, D. Effective charge transport in poly (3,4-ethylenedioxythiophene) based hybrid films containing polyoxometallate redox centers. *J. Electrochem. Soc.* **2005**, *152*, E98. [CrossRef]
61. Szymanska, D.; Rutkowska, I.A.; Adamczyk, L.; Zoladek, S.; Kulesza, P.J. Effective charge propagation and storage in hybrid films of tungsten oxide and poly (3, 4-ethylenedioxythiophene). *J. Solid State Electrochem.* **2010**, *14*, 2049–2056. [CrossRef]
62. Adamczyk, L. Development of copper-stabilized conducting-polymer/polyoxometalate hybrid materials for effective electrochemical charging. *J. Solid State Electrochem.* **2017**, *21*, 211–222. [CrossRef]
63. Adamczyk, L.; Kulesza, P.J. Fabrication of composite coatings of 4-(pyrrole-1-yl) benzoate-modified poly-3,4-ethylenedioxythiophene with phosphomolybdate and their application in corrosion protection. *Electrochim. Acta* **2011**, *56*, 3649–3655. [CrossRef]
64. de Oliveira, A.d.N.; Barbosa de Lima, M.A.; de Oliveira Pires, L.H.; Rosas da Silva, M.; Souza da Luz, P.T.; Angélica, R.S.; da Rocha Filho, G.N.; da Costa, C.E.F.; Luque, R.; Santos do Nascimento, L.A. Bentonites Modified with Phosphomolybdic Heteropolyacid (HPMo) for Biowaste to Biofuel Production. *Materials* **2019**, *12*, 1431. [CrossRef]
65. da Silva, M.J.; de Oliveira, C.M. Catalysis by Keggin heteropolyacid salts. *Curr. Catal.* **2018**, *7*, 26–34. [CrossRef]
66. Adamczyk, L.M.; Kulesza, P.J. Preparation and Protective Properties of Composite Films of Poly (3,4-etylenedioxythiophene) with Seteropolyanions on Stainless Steel. *ECS Trans.* **2008**, *13*, 85. [CrossRef]
67. Kaoua, R.; Bennamane, N.; Bakhta, S.; Benadji, S.; Rabia, C.; Nedjar-Kolli, B. Synthesis of Substituted 1,4-Diazepines and 1,5-Benzodiazepines Using an Efficient Heteropolyacid-Catalyzed Procedure. *Molecules* **2011**, *16*, 92–99. [CrossRef]
68. Adamczyk, L.; Miecznikowski, K. Solid-state electrochemical behavior of Keggin-type borotungstic acid single crystal. *J. Solid State Electrochem.* **2013**, *17*, 1167–1173. [CrossRef]
69. Song, I.K.; Barteau, M.A. Redox properties of Keggin-type heteropolyacid (HPA) catalysts: Effect of counter-cation, heteroatom, and polyatom substitution. *J. Mol. Catal. A Chem.* **2004**, *212*, 229–236. [CrossRef]
70. Han, D.S.; Hwang, T. An adaptive pre-distorter for the compensation of HPA nonlinearity. *IEEE Trans. Broadcast.* **2000**, *46*, 152–157.
71. Song, I.K.; Barteau, M.A. Scanning tunneling microscopy (STM) and tunneling spectroscopy (TS) of heteropolyacid (HPA) self-assembled monolayers (SAMS): Connecting nano properties to bulk properties. *Korean J. Chem. Eng.* **2002**, *19*, 567–573. [CrossRef]

72. Park, D.R.; Park, S.; Bang, Y.; Song, I.K. Redox properties and oxidation catalysis of group 5 metal (V, Nb, Ta)-containing Keggin and Wells–Dawson heteropolyacid (HPA) catalysts. *Appl. Catal. A Gen.* **2010**, *373*, 201–207. [CrossRef]
73. Bidan, G.; Genies, E.M.; Lapkowski, M. Polypyrrole and poly (N-methylpyrrole) films doped with Keggin-type heteropolyanions: Preparation and properties. *J. Electroanal. Chem. Interfacial Electrochem.* **1988**, *251*, 297–306. [CrossRef]
74. Saenko, E.; Sarafanov, A.; Ananyeva, N.; Behre, E.; Shima, M.; Schwinn, H.; Josić, D. Comparison of the properties of phospholipid surfaces formed on HPA and L1 biosensor chips for the binding of the coagulation factor VIII. *J. Chromatogr. A* **2001**, *921*, 49–56. [CrossRef]
75. White, A.M.; Slade, R.C. Electrochemically and vapour grown electrode coatings of poly (3,4-ethylenedioxythiophene) doped with heteropolyacids. *Electrochim. Acta* **2004**, *49*, 861–865. [CrossRef]
76. Adekoya, G.J.; Sadiku, R.E.; Ray, S.S. Nanocomposites of PEDOT: PSS with graphene and its derivatives for flexible electronic applications: A review. *Macromol. Mater. Eng.* **2021**, *306*, 2000716. [CrossRef]
77. McDonald, M.B.; Freund, M.S. Novel conducting polymer-heteropoly acid hybrid material for artificial photosynthetic membranes. *ACS Appl. Mater. Interfaces* **2011**, *3*, 1003–1008. [CrossRef]
78. Góral, M.; Jouini, M.; Perruchot, C.; Miecznikowski, K.; Rutkowska, I.A.; Kulesza, P.J. Integration of vanadium-mixed addenda Dawson heteropolytungstate within poly (3,4-ethylenedioxythiophene) and poly (2,2′-bithiophene) films by electrodeposition from the nonionic micellar aqueous medium. *Electrochim. Acta* **2011**, *56*, 3605–3615. [CrossRef]
79. Badea, G.E.; Fodor, A.; Petrehele, A.I.G.; Maior, I.; Toderaș, M.; Morgovan, C.M. Evaluation of phosphopolyoxometalates with mixed Addenda (Mo, W, V) as corrosion inhibitors for steels. *Materials* **2023**, *16*, 7600. [CrossRef] [PubMed]
80. Wang, X.F.; Liu, X.Y.; Su, F.; Li, J.S.; Zhu, Z.M.; Sang, X.J.; Zhang, L.C. Enhanced corrosion resistance of carbon steel in hydrochloric acid solution by Polyoxometalate-Estertin derivatives. *ACS Omega* **2022**, *7*, 4429–4443. [CrossRef] [PubMed]
81. Romaric Wamba Tchio, O.; Pengou, M.; Baumier, C.; Franger, S.; Teillout, A.L.; Martyr Mbomekallé, I.; Ngameni, E. Comparison between Lacunary and Saturated Keggin Polyoxometalates as Steel Corrosion Inhibitors in Chloride Solution: Contribution of the Lacuna in the Inhibition Mechanism. *ChemistrySelect* **2020**, *5*, 10135–10143. [CrossRef]
82. Standard Test Methods for Measuring Adhesion by Tape Test. Available online: https://www.galvanizeit.com/uploads/resources/ASTM-D-3359-yr-2010.pdf (accessed on 12 December 2024).
83. Kucharczyk, A.; Adamczyk, L.; Miecznikowski, K. The Influence of the Type of Electrolyte in the Modifying Solution on the Protective Properties of Vinyltrimethoysilane/Ethanol-Based Coatings Formed on Stainless Steel X20Cr13. *Materials* **2021**, *14*, 6209. [CrossRef]

materials

MDPI

Article

Preliminary Aspects Regarding the Anticorrosive Effect of Multi-Layered Silane–Hydroxyapatite Coatings Deposited on Titanium Grade 2 for Medical Applications

Agata Dudek * and Oliwia Kierat *

Department of Material Engineering, Faculty of Production Engineering and Materials Technology, Czestochowa University of Technology, Aleja Armii Krajowej 19, 42-200 Czestochowa, Poland
* Correspondence: agata.dudek@pcz.pl (A.D.); oliwia.kierat@pcz.pl (O.K.)

Abstract: This paper presents a method for producing VTMS/HAp/VTMS/VTMS multilayer coatings on a Grade 2 titanium substrate and characterizes their structure and functional properties. Two solutions were used to produce the coatings: one based on vinyltrimethoxysilane (VTMS) and the other on hydroxyapatite (HAp) powder. The coatings were applied using immersion using the sol-gel method. Microstructural tests of the multilayer coatings were performed, their chemical composition was determined, and the structure was characterized using Fourier Transform Infrared Spectroscopy (FTIR). A detailed analysis of the geometric structure of the coatings was carried out both before and after corrosion tests. The geometric structure of the multilayer coatings was analyzed using a light microscope and an atomic force microscope (AFM). The thickness of the coatings was determined using a Testan DT-10 AN 120 157 m, and the adhesion of the coatings to the substrate was analyzed using Scotch™ tape. The corrosion resistance of the coatings in simulated body fluid was tested to evaluate their suitability for implantology. As demonstrated by the research presented in this paper, the sol–gel process can successfully produce silane coatings by adding hydroxyapatite powder. The new materials proposed in this study can effectively protect metal materials used in medicine against corrosion.

Keywords: titanium; surface modification; sol–gel method; silane coatings; hydroxyapatite coatings

1. Introduction

Surface modification of biomaterials is a rapidly developing area of research in many scientific centers that aim to improve the properties of materials used in medicine. The proposed techniques focus on surface modification to better meet the specific requirements of their intended applications. Metallic biomaterials, widely used in the production of implants, despite their numerous advantages, have certain limitations. For example, they are often too stiff compared to natural bone, can degrade over time, which affects their stability and durability, and are exposed to corrosion induced by the aggressive physiological environment.

Titanium and its alloys have gained particular recognition among the metallic materials used for implants. They are characterized by excellent chemical and mechanical properties, high biocompatibility, excellent corrosion resistance, low specific gravity, and modulus of elasticity [1–9]. To extend the life of implants, thin-layer or multilayer coatings are increasingly recommended, as these can provide protection, including effective corrosion resistance [10–13].

Changes to the surface layer lead to improved biocompatibility, corrosion resistance, and wear resistance [14]. A literature analysis shows that numerous studies confirm the benefits of surface modification of titanium and its alloys in the context of implantology [15–19]. Surface engineering methods can be divided into three main categories: (i) mechanical methods, (ii) chemical methods, and (iii) physical methods. Figure 1. presents methods for modifying the surface of titanium and its alloys. The figure presents various surface modification methods of titanium and its alloys, identified based on a review of numerous scientific publications. These methods aim to improve the physical, chemical, and mechanical properties and tailor its surface for specific applications, especially in the medical industry.

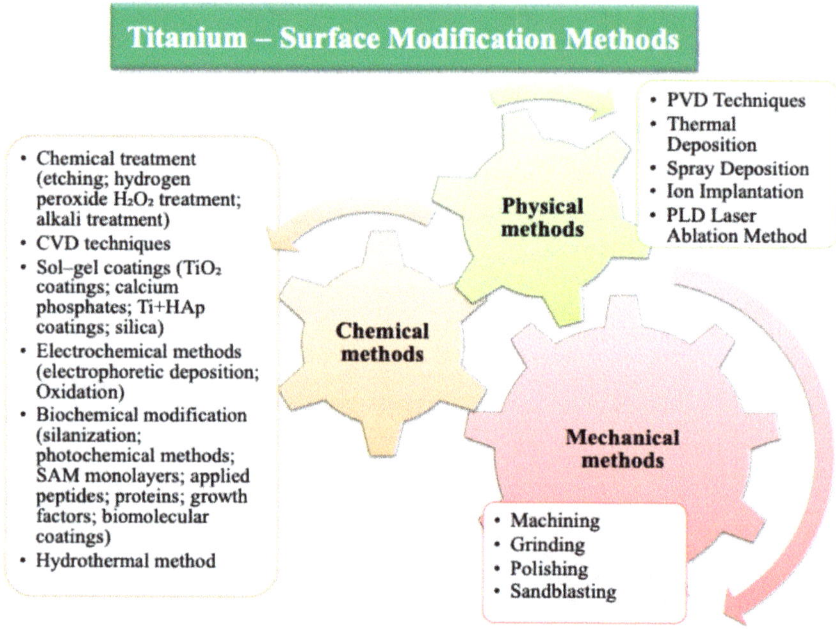

Figure 1. Methods of modifying the surface of titanium and its alloys—own study based on [15,20–26].

Sol–gel methods are widely used in medicine to produce coatings on titanium and its alloys. This leads to the creation of a new generation of biomaterials with precisely defined microstructure, chemical composition, and surface topography. Moreover, sols can be enriched with substrates that impart desired properties to the coating, such as biocompatibility or antithrombogenicity.

In this study, the production of multilayer coatings consisting of four layers, applied from two solutions—based on hydroxyapatite and vinyltrimethoxysilane—is proposed. The purpose of using hydroxyapatite coatings on metal biomaterials is to limit the release of metal ions from the substrate and to increase the bioactivity of bone tissue [27,28]. Despite the numerous advantages associated with the use of hydroxyapatite-based coatings in implantology, there are also limitations, such as their high brittleness and low fracture resistance [29].

The use of vinyltrimethoxysilane as a silicon-based coupling agent is intended to improve the performance of hydroxyapatite coatings. As part of this work, a detailed analysis of the structure and performance of VTMS/HAp/VTMS/VTMS coatings was produced and performed, including tests of thickness, adhesion of coatings to the substrate, microstructure, chemical composition, geometric structure, and corrosion resistance. In the literature, many studies address surface modifications of implants; however, to date, very few studies have investigated the use of silane-based coatings with the addition of hydroxyapatite (HAp) as an innovative material solution. The coating we propose is a new and unique approach in this field, combining the properties of silane and HAp, which can significantly improve osseointegration and implant durability. Therefore, highlighting this novel coating in the context of existing surface modification methods is of scientific importance. The results of our study confirm the excellent performance characteristics of this coating, making a valuable contribution to the development of technologies used in implantology and opening up new possibilities for clinical practice.

2. Materials and Methods

In this work, titanium Grade 2 was selected as the substrate for the produced coatings with the following chemical composition: Fe—max. 0.3; O—max. 0.25; C—max. 0.08; N—max. 0.03; H—max. 0.015, Ti—the rest. Samples for corrosion tests were in the form of cylinders with a diameter of 5 mm, which were mounted in polymethyl methacrylate frames using epoxy resin. For the remaining tests, rods with a diameter of 25 mm and a height of about 5 mm were used. The coating procedure was as follows: (i) wet mechanical polishing of the samples using 2000 grit sandpaper, (ii) washing of the samples with distilled water, (iii) degreasing of the samples with acetone, (iv) immersion of the samples in the prepared solutions for the appropriate time, (v) removal of excess solution using filter paper, and (vi) storage of the coated samples in a desiccator until they were completely dry. The coatings were prepared using analytically pure reagents, namely the following: (i) vinyltrimethoxysilane (VTMS) from Sigma Aldrich (St. Louis, MO, USA), (ii) hydroxyapatite powder (HAp) from Sulzer Metco (Winterthur, Switzerland), (iii) anhydrous ethyl alcohol (EtOH) from Chempur (Piekary Slaskie, Poland), and (iv) acetic acid (AcOH) from Chempur. The coatings were prepared by preparing and applying two solutions: (i) a vinyltrimethoxysilane-based solution and (ii) a hydroxyapatite-based solution. The silane-based solution was obtained by mixing vinyltrimethoxysilane, ethyl alcohol, acetic acid, and distilled water according to the procedure developed in earlier works [30,31]. The volume ratio of VTMS:EtOH:AcOH: H_2O was 0.6:0.2:0.06:0.14. In turn, the hydroxyapatite-based solution was obtained by mixing 5 g of hydroxyapatite powder and 2.5 g of ethanol. Both solutions were mixed using a magnetic stirrer. The mixing time of the vinyltrimethoxysilane-based solution was 24–72 h, while the hydroxyapatite-based solution was 1–5 h. The produced coatings consisted of 4 layers—details regarding the formation of the coatings are included in patent application no. P.440129.

The hydroxyapatite powder used in the study has an average particle diameter of 19.93 μm, with a maximum diameter of 87.07 μm and a minimum of 5.34 μm. Figure 2 shows the microstructure of the hydroxyapatite powder, as observed using digital microscopy.

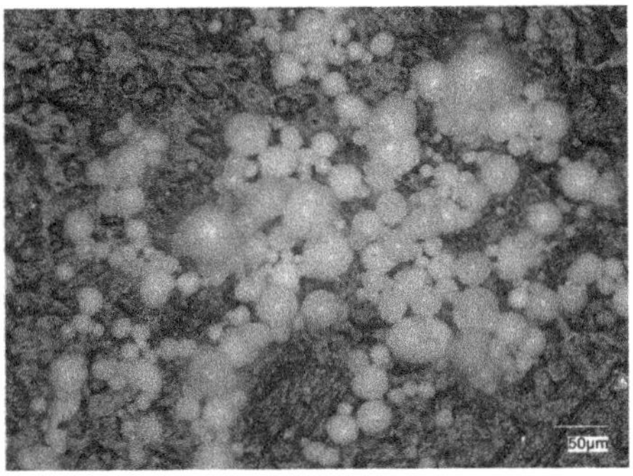

Figure 2. Microstructure of the hydroxyapatite powder used in the study, recorded with a KEYENCE VHX-7000 (Keyence, Mechelen, Belgium) digital microscope.

The adhesion of the coatings to the substrate was tested using Scotch™ tape (Scotch™ Brand, St. Paul, MN, USA) with the pull-off method according to ASTM D3359 standard [32]. The coating thickness was measured using a Testan DT-10 AN 120 157 (Anticorr, Gdansk, Poland) thickness gauge (Anticorr, Gdansk, Poland). The measurement consisted of establishing a calibration point on an uncoated substrate, measuring the coating thickness 10 times, and determining the average value of the coating thickness.

Microstructural studies were performed using a JEOL JSM-6610 LV scanning electron microscope (Jeol, Tokyo, Japan) and a KEYENCE VHX-7000 digital microscope (Keyence, Mechelen, Belgium). The chemical composition of the coatings was characterized using an EDS (Energy Dispersive Spectroscopy) X-ray microanalyzer—X-Max 20mm^2 (Oxford Instruments, Abingdon, Oxfordshire, UK) attached to a scanning electron microscope (SEM). The structure of the produced coating was characterized using an IRAffinity-1S FTIR SHIMADZU infrared spectrophotometer (Kyoto, Japan) with a QATR 10 attachment. The roughness of the coatings was determined using a KEYENCE VHX-7000 digital microscope and a NanoScope V MultiMode 8 Bruker atomic force microscope (Bruker, Bremen, Germany), which was also used to measure the topography of the coatings. During the AFM measurements, the tapping mode was used with a scan rate of 1 Hz. The characteristics of the cantilever used in the measurements were as follows: Bruker, model—TESP; material: 0.01–0.025 Ohm-cm Antimony (n) doped Di; f: 305–368 kHz; K: 20–80 N/m.

To determine the surface roughness, both two-dimensional and three-dimensional surface analyses were taken into account. In the two-dimensional analysis of surface irregularities, the following parameters were taken into account to describe the roughness: the arithmetic mean of the roughness profile deviations (Ra), the maximum height of the roughness profile (Rz), the ten-point irregularity height (RzJIS) which is the sum of the average absolute values of the five highest-profile peaks and the average absolute values of the five lowest profile depressions measured with respect to the mean line of the elementary section of the roughness profile, the maximum height of the roughness profile peak (Rp), the maximum depth of the roughness profile depression (Rv), the average height of the roughness profile elements (Rc), the total height of the roughness profile (Rt), the root mean square of the roughness profile deviations (Rq), the roughness profile asymmetry coefficient—skewness (Rsk), the roughness profile slope coefficient—kurtosis (Rku), the average width of the roughness profile elements (RSm), and the mean square elevation of the roughness profile (Rdq).

Taking into account the three-dimensional analysis of surface irregularities, the following parameters were used to describe them: the arithmetic mean deviation of the surface from the mean surface (Sa), the maximum surface height (Sz), the standard deviation of the surface irregularities (Sq), the surface asymmetry coefficient—skewness (Ssk), the surface inclination coefficient—kurtosis (Sku), the highest surface elevation (Sp), and the depth of the lowest surface depression (Sv).

The corrosion resistance tests were performed using the CH Instruments 440A measuring station (CH Instruments, Austin, TX, USA) in a three-electrode system: (i) an auxiliary electrode—platinum, (ii) a reference electrode—a saturated calomel electrode (SCE), and (iii) the tested electrode. For each sample, a potentiodynamic curve was recorded in the potential range from −1.5 V to 3.0 V.

The prepared samples were analyzed in simulated body fluid, the composition of which is presented in Table 1.

Table 1. Chemical composition of simulated body fluid [33].

SBF, Simulated Body Fluid	
NaCl	8.035 g dm^{-3}
KCl	0.225 g dm^{-3}
$CaCl_2$	0.292 g dm^{-3}
$NaHCO_3$	0.355 g dm^{-3}
$MgCl_2 \cdot 6H_2O$	0.311 g dm^{-3}
$K_2HPO_4 \cdot 3H_2O$	0.231 g dm^{-3}
Na_2SO_4	0.072 g dm^{-3}
$((HOCH_2)_3CNH_2)$	6.118 g dm^{-3}
HCl (1 mol dm^{-3})	39 mL dm^{-3}

Many solutions simulate the conditions prevailing in the human body, the so-called simulated physiological solutions, among which SBF is characterized by the largest number of components and the highest concentration of chloride ions in comparison to other solutions used, among others, in the work [31].

3. Results and Discussion

3.1. Peeling Test of VTMS/HAp/VTMS/VTMS Coating

The results showed very good adhesion of the coatings to the base material. After five times of sticking and peeling the tape from the surface, no visible damage or defects were found, which indicates strong adhesion. After each tapping and unsticking cycle, the surface was microscopically analyzed in detail.

Analysis of these results suggests that the presented coatings have an exceptional ability to maintain integrity in contact with the titanium substrate, which may be crucial in the context of their use in medical environments, where durability and reliability of corrosion protection are particularly important. Such results provide a solid basis for further development and potential implementation of these materials in medical practice, especially in the case of implants and other devices in contact with body tissues.

The pull-off test was used as a preliminary method to evaluate the adhesion of the coatings. In the future, we plan to conduct more advanced measurements to provide detailed, quantitative information on the adhesion of the coatings.

3.2. Thickness of the Produced Coatings

The average thickness of the VTMS/HAp/VTMS/VTMS coating deposited on a Grade 2 titanium substrate was 5.27 μm. Thickness analysis allows for the assessment of the efficiency of the coating production process and its potential functionality in medical applications. This value indicates the appropriate thickness, which provides not only effective protection against corrosion but also is optimal, allowing the coating to cooperate with the titanium substrate in dynamic physiological conditions. The thickness of the coating in this range can affect its resistance to mechanical damage and the ability to maintain chemical stability in long-term use, which is important in the context of its potential use in implantology and other medical devices.

3.3. Microstructural Studies of the VTMS/HAp/VTMS/VTMS Coating

A detailed analysis of the morphology of the produced coatings was carried out. Figure 3 shows the microstructure of the VTMS/HAp/VTMS/VTMS coating deposited on Grade 2 titanium. Based on the recorded images, it was found that the coating was evenly distributed on the substrate surface, showing no signs of cracks or unevenness. Its structure was compact and uniform, and the coating thickness was uniform, without visible fluctuations.

The particles present in the coating structure, visible in the microscopic images, are related to the presence of hydroxyapatite (HAp) in powder form, which confirms the effective incorporation of this substance into the coating structure.

The morphological analysis indicates the high quality of the produced coating, which may be of significant importance in the context of its medical applications, especially in the case of implants, where homogeneity, lack of cracks, and an appropriate surface structure are key to ensuring long-term durability and effectiveness of corrosion protection. The homogeneity of the coating may also promote better integration of the material with body tissues, which is important from the point of view of biocompatibility. The high quality of the produced coating applied using the sol–gel method refers to key properties such as uniformity and thickness control of the coating, good adhesion to the substrate, and corrosion resistance, which determine their effectiveness and durability in various applications.

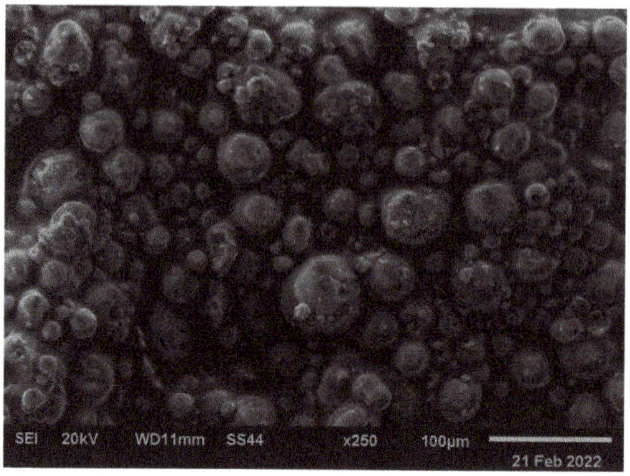

Figure 3. The microstructure of the VTMS/HAp/VTMS/VTMS coating deposited on Grade 2 titanium was imaged using a scanning electron microscope.

3.4. Chemical Composition of the VTMS/HAp/VTMS/VTMS Coating

A detailed analysis of the chemical composition of the multilayer coating deposited on a Grade 2 titanium substrate was carried out. As a result of this analysis, the presence of several key elements, C, O, Si, P, and Ca, was revealed, which are an integral part of the coating structure. Figure 4 shows the EDS spectrum, which allowed the identification of these elements, as well as the determination of their content in VTMS/Hap/VTMS/VTMS coating. The measurement results also include measurement uncertainties expressed in weight percent.

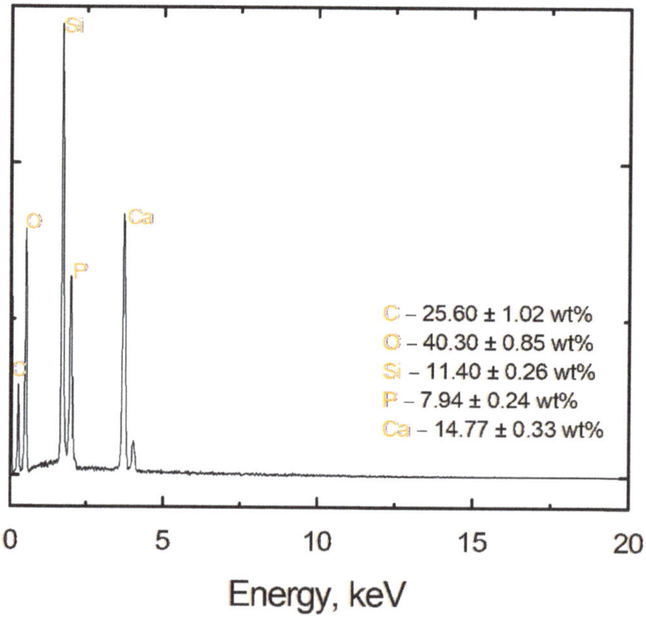

Figure 4. EDS spectrum recorded for the VTMS/HAp/VTMS/VTMS coating deposited on Grade 2 titanium, showing the distribution of individual elements.

Analysis of the chemical composition of the coating provided important information on the structure and potential functional properties of the coatings. The EDS analysis results confirm that the coating consists of appropriate components that can ensure its durability, corrosion resistance, and beneficial biological interactions in medical applications.

3.5. Characteristics of the VTMS/HAp/VTMS/VTMS Coating Structure

The spectrum recorded for the VTMS/HAp/VTMS/VTMS coating deposited on Grade 2 titanium, presented in Figure 5, reveals the presence of characteristic bands in different wavenumber ranges, which indicate specific functional groups included in the coating. In addition to the spectrum for the multilayer coating, a spectrum was also recorded for hydroxyapatite powder. The spectrum obtained for the VTMS/HAp/VTMS/VTMS coating shows bands for wavenumbers around 400 cm^{-1}, which originate from Ti-O-Ti stretching vibrations [34]. The bands in the range of 500–660 cm^{-1} indicate the presence of the PO_4 group in hydroxyapatite. The peak recorded for the wave number of 753 cm^{-1} is related to the C-H bond, while for the wave number of 887 cm^{-1}, it most probably comes from the Si-O-Ti bond. The bands in the range of 930–1220 cm^{-1} result from both the stretching of the Si-O bond and the presence of PO_4, which is confirmed by, among others, the spectrum recorded for the hydroxyapatite powder itself [35,36]. The peak recorded for the wave number of 1273 cm^{-1} occurs due to the bending of the C-H bond, while the peaks at 1412 and 1602 cm^{-1} are related to the unconjugated C=C bond [37]. Characteristic C-H stretching bands occur at wave numbers 2844, 2959, and 3061 cm^{-1} [35].

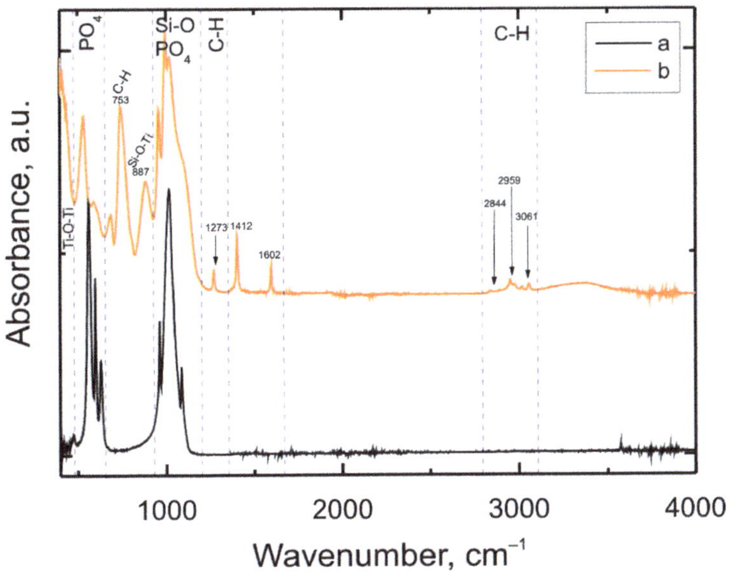

Figure 5. FTIR spectra were recorded for hydroxyapatite powder (a) and the VTMS/HAp/VTMS/VTMS coating deposited on Grade 2 titanium (b).

Based on the analysis of the spectra obtained for the VTMS-HAp-VTMS-VTMS coating, the occurrence of absorbance maxima for the functional groups Ti-O-Ti, Si-O, C-H, C=C, Si-O-Ti, PO_4^{3-} can be observed.

The presence of characteristic bands for Ti-O-Ti, PO_4, Si-O-Ti, and C-H confirms the effective deposition of VTMS and HAp layers on the titanium surface and their mutual interactions. The obtained chemical composition suggests that the coating may exhibit bioactive properties, which is particularly important in the context of medical applications such as implants. The bioactivity of the coating, due to the presence of hydroxyapatite, can support bone integration, while the silanes contribute to the stability and durability of the coating, ensuring its resistance to mechanical and chemical factors.

3.6. Geometric Structure of the VTMS/HAp/VTMS/VTMS Coating

3.6.1. Roughness of the Grade 2 Titanium

Figure 6 shows the microstructure and surface topography of Grade 2 pure titanium obtained using a digital microscope. By capturing isometric images and control maps with high sharpness and resolution, 2D profile roughness parameters and surface roughness parameters were determined for the substrate used for the coating application. Table 2 presents the 2D profile roughness parameters, while Table 3 presents the surface roughness parameters determined for Grade 2 titanium.

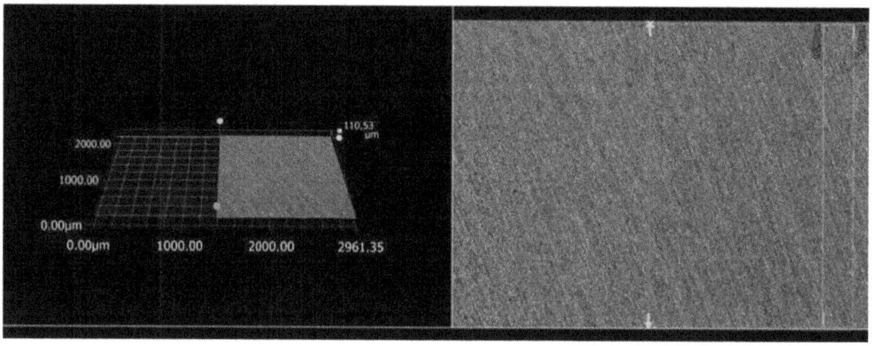

Figure 6. Line roughness was measured using a KEYENCE VHX-7000 digital microscope for a Grade 2 titanium.

Table 2. Two-dimensional profile roughness parameters for a Grade 2 titanium.

Ti Grade 2	
Ra [μm]	17.74
Rz [μm]	70.82
RzJIS [μm]	0
Rp [μm]	32.69
Rv [μm]	38.12
Rt [μm]	70.86
Rq [μm]	20.59
Rsk	−0.16
Rku	1.84
Rdq	0.04

Table 3. Surface roughness parameters for a Grade 2 titanium.

Ti Grade 2	
Sa [μm]	21.07
Sz [μm]	108.67
Sq [μm]	24.64
Ssk	−0.14
Sku	1.99
Sp [μm]	47.23
Sv [μm]	61.43

3.6.2. Roughness of the VTMS/HAp/VTMS/VTMS Coating Before Corrosion Tests

Figure 7 presents the microstructure and surface topography of the tested coatings obtained on a digital microscope, which was performed to determine the 2D profile roughness and surface roughness parameters. The obtained isometric images and contour maps were characterized by high sharpness, contrast, and resolution. Based on the measurements performed, information on the profile, its height, and stereometric parameters was obtained.

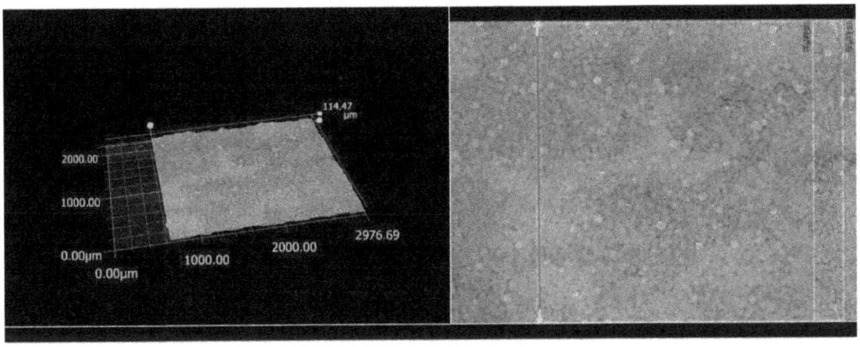

Figure 7. Line roughness was measured using a KEYENCE VHX-7000 digital microscope for a VTMS/HAp/VTMS/VTMS coating deposited on a Grade 2 titanium substrate.

The 2D profile roughness and surface roughness parameters obtained for the tested coatings are presented in Tables 4 and 5.

Table 4. Two-dimensional profile roughness parameters for a VTMS/HAp/VTMS/VTMS coating deposited on a Grade 2 titanium substrate.

VTMS/HAp/VTMS/VTMS Coating	
Ra [µm]	5.74
Rz [µm]	43.72
RzJIS [µm]	17.19
Rp [µm]	26.84
Rv [µm]	16.88
Rc [µm]	23.35
Rt [µm]	43.72
Rq [µm]	7.98
Rsk	0.94
Rku	5.09
RSm [µm]	693.91
Rdq	0.31

Table 5. Surface roughness parameters for VTMS/HAp/VTMS/VTMS coating deposited on a Grade 2 titanium substrate.

VTMS/HAp/VTMS/VTMS Coating	
Sa [µm]	11.35
Sz [µm]	112.68
Sq [µm]	14.74
Ssk	0.02
Sku	3.58
Sp [µm]	68.88
Sv [µm]	43.8

3.6.3. Roughness of the VTMS/HAp/VTMS/VTMS Coating After Corrosion Tests

Assessment of surface topography was also performed for coatings after corrosion resistance tests. This analysis aimed to check the effect of the corrosive solution—in this case, simulated body fluid—on the geometric structure of the coating. Figure 8 presents the microstructure and surface topography of the tested coatings obtained on a digital microscope, which was performed to determine the 2D profile roughness and surface roughness parameters. Tables 6 and 7 present the two-dimensional and three-dimensional roughness parameters for the coating after corrosion tests.

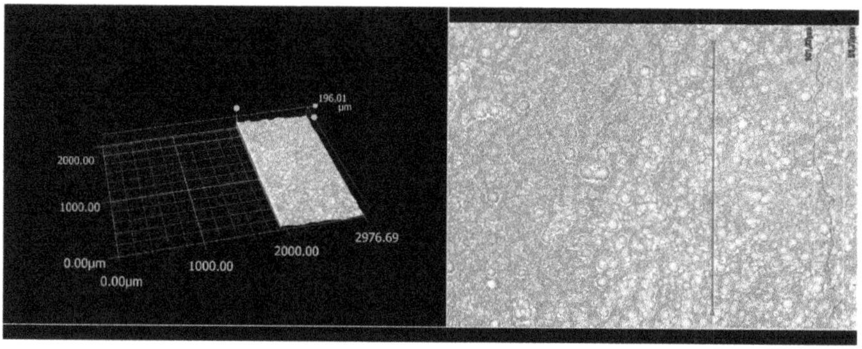

Figure 8. Line roughness was measured using a KEYENCE VHX-7000 digital microscope for a VTMS/HAp/VTMS/VTMS coating deposited on a Grade 2 titanium substrate after corrosion testing in simulated body fluid.

Table 6. Two-dimensional profile roughness parameters for a VTMS/HAp/VTMS/VTMS coating deposited on a Grade 2 titanium substrate after corrosion testing in simulated body fluid.

VTMS/HAp/VTMS/VTMS Coating After Corrosion Tests	
Ra [μm]	12.22
Rz [μm]	49.57
RzJIS [μm]	29.86
Rp [μm]	20.19
Rv [μm]	29.38
Rc [μm]	29.86
Rt [μm]	49.57
Rq [μm]	13.57
Rsk	−0.07
Rku	1.85
RSm [μm]	548.55
Rdq	0.16

Table 7. Surface roughness parameters for VTMS/HAp/VTMS/VTMS coating deposited on a Grade 2 titanium substrate after corrosion testing in simulated body fluid.

VTMS/HAp/VTMS/VTMS Coating After Corrosion Tests	
Sa [μm]	26.86
Sz [μm]	190.19
Sq [μm]	33.56
Ssk	−0.03
Sku	2.72
Sp [μm]	99.03
Sv [μm]	91.16

Analyzing the obtained isometric images and contour maps, it can be seen that the morphology and topography of the coatings differ from each other. Based on the analysis of the surface roughness parameters, both in the context of the profile and topography of the coatings, it can be concluded that the exposure of the coating to the corrosive solution leads to an increase in the surface roughness.

Literature data show that bioactive coatings, such as hydroxyapatite (HAp) or titanium anti-corrosion coatings, can reduce the corrosion rate of the implant material and, at the same time, improve osteointegration due to adequate surface roughness. Many authors have highlighted the benefits of a surface with the appropriate level of roughness for the primary stability and osseointegration process of implants, influencing both mechanical properties and cellular response. However, despite decades of research, the impact of implant surface roughness and other surface parameters on the biological response remains a complex and controversial topic. Some studies also point to opposing trends, suggesting that certain levels of roughness may negatively affect osteoblast behavior or lead to different outcomes in the osseointegration process. To date, no clear consensus has been reached on which surface parameters reliably and definitively predict osteoblast behavior on titanium and titanium alloy implant surfaces [38,39].

3.7. Topography of the Produced Multilayer Coatings

Figure 9 shows the 2D and 3D topography image and the roughness profile recorded for the VTMS/Hap/VTMS/VTMS coating deposited on Grade 2 titanium. The roughness of the multilayer coatings recorded using the atomic force microscope was 0.366 nm. The differences in roughness obtained using the atomic force microscope and the digital microscope result from the analysis of surfaces of different sizes. The area examined with an atomic force microscope was only 1×1 µm, which may affect the accuracy and detail of the measurements obtained.

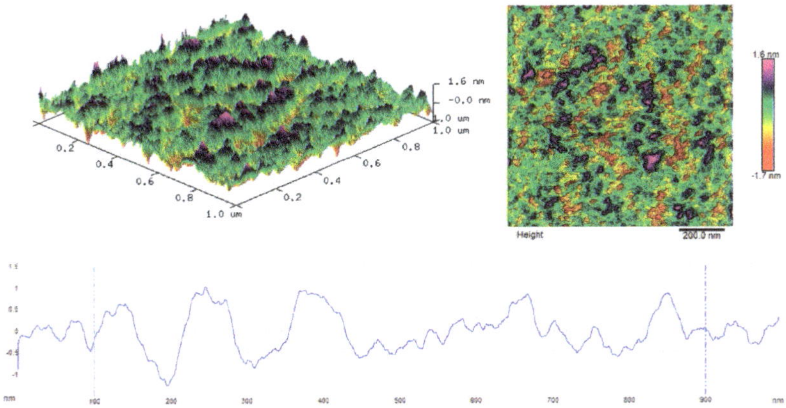

Figure 9. Three-dimensional and two-dimensional topography and roughness profile of the VTMS/Hap/VTMS/VTMS coating deposited on Grade 2 titanium.

3.8. Corrosion Resistance Tests

Figure 10 shows the potentiodynamic polarization curves recorded for (a) Grade 2 titanium and (b) the VTMS/HAp/VTMS/VTMS coating deposited on Grade 2 titanium (A) and a photo of the coating after corrosion tests in simulated body fluid (B), which confirms the lack of coating damage as a result of corrosion resistance tests.

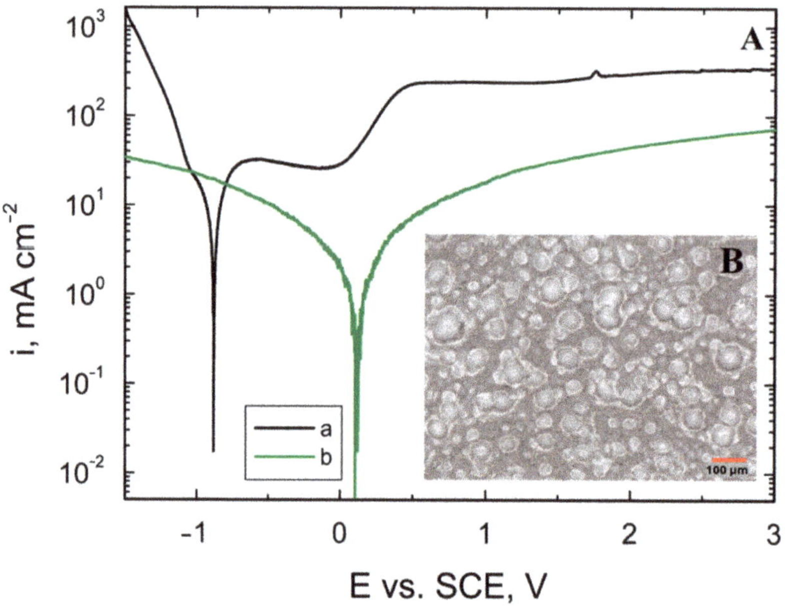

Figure 10. Potentiodynamic polarization curves (**A**) were analyzed in simulated body fluid for (a) Grade 2 titanium and (b) the coating deposited on Grade 2 titanium, along with the coating microstructure after corrosion tests in simulated body fluid (**B**).

Based on the recorded potentiodynamic polarization curves, the polarization resistances were determined for a clean Grade 2 titanium substrate (a) and a VTMS/HAp/VTMS/VTMS coating deposited on Grade 2 titanium (b). Potentiodynamic measurements were performed with log(i) = f(E), and the curves were recorded at a polarization rate of 0.01 V/s. The relatively high scan rate may have limited the depth of alloy and coating etching during a single measurement. However, it was sufficient to record Faradaic processes on the electrode. The total duration of a single measurement was 7.5 min. In the future, we will also consider the use of gravimetric and impedance tests to determine precise values of corrosion rates.

Figure 11 shows linear graphs of the polarization resistance for Grade 2 titanium and a VTMS/HAp/VTMS/VTMS coating deposited on Grade 2 titanium. Linear polarization resistance plots show the dependence $\Delta U = f(\Delta i)$ near E_{corr}. For potentials slightly different from E_{corr} ($\pm$25 mV), the external current density is a linear function of the potential, while the slope of the corresponding lines is a measure of the polarization resistance, which confirms the Stern–Hoar equation [40]. Table 8 shows the values of the corrosion potential and polarization resistance for a clean Grade 2 titanium substrate (a) and a VTMS/HAp/VTMS/VTMS coating deposited on a Grade 2 titanium substrate (b).

Table 8. Summary of parameters obtained for individual samples.

Sample	Corrosion Potential [V]	Polarization Resistance [kΩ cm^2]
Ti Gr2	*−0.88*	*4.862 ± 0.038*
Ti Gr2 + coating	*0.10*	*34.267 ± 2.761*

As is known, the higher the value of the corrosion potential (E_{corr}) and the lower the values of the anodic currents, the higher the resistance to the corrosive environment of the material.

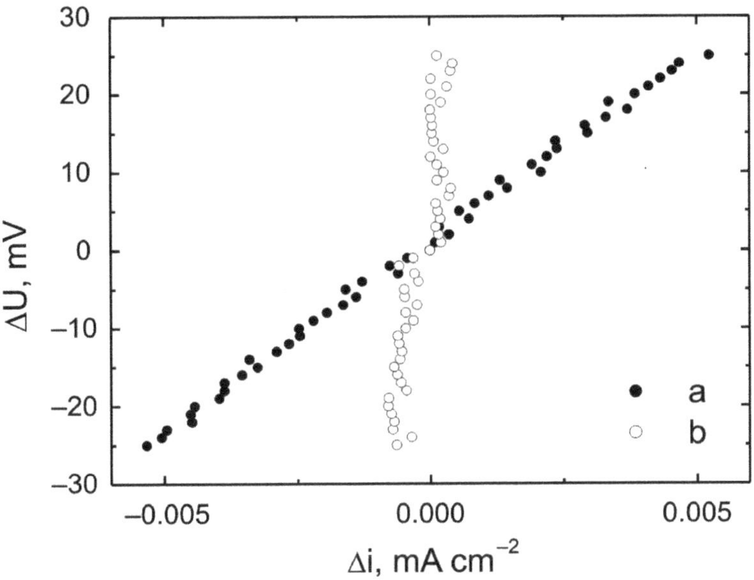

Figure 11. Linear polarization resistance plots were recorded for (a) Grade 2 titanium and (b) the coating deposited on Grade 2 titanium in simulated body fluid.

The analysis of the presented potentiodynamic curves and polarization resistance measurements in the simulated body fluid show that the best results were achieved for the titanium alloy covered with the VTMS-HAp-VTMS-VTMS coating (Figure 10b). The use of this coating on the titanium surface increases the resistance of the tested material to corrosion, providing effective anti-corrosion protection. The VTMS-HAp-VTMS-VTMS coating on the titanium substrate significantly reduced both the cathodic and anodic current densities compared to the material without the coating. Additionally, a significant shift in the corrosion potential (E_{corr}) towards positive values was observed, which is illustrated by the change from $E_{corr} = -0.88$ V for titanium without the coating to $E_{corr} = 0.10$ V for the Grade 2 titanium with the VTMS-HAp-VTMS-VTMS coating. The highest polarization resistance value was obtained for the VTMS-HAp-VTMS-VTMS coating, which confirms high corrosion resistance.

Silane-based sol–gel coatings can be used to protect metals and alloys as an alternative to toxic chromate conversion coatings [41]. Silane coatings can act as primers by creating a physical barrier that restricts the access of corrosive agents to the metal surface, thus protecting against corrosion. The effectiveness of silane coatings relies on the formation of a defect-free layer with strong adhesion. Several factors influence the formation of silane coatings, including curing conditions, solution pH, surface preparation, coating thickness, application method, and particle size matrix [42–44].

4. Conclusions

High potential of the new material solution: The results of the conducted tests confirm the high potential of the proposed material solution, which can provide effective corrosion protection for metallic materials used in medical applications, especially in implantology.

Effectiveness of the applied coating: The VTMS-HAp-VTMS-VTMS multilayer coating, produced by the sol–gel method using VTMS silane and hydroxyapatite (HAp) powder, demonstrates excellent adhesion to the substrate and effectively protects the titanium substrate from the corrosive environment. Its thickness is approximately 5.27 μm, and its roughness values are Ra = 5.74 μm and Sa = 11.35 μm, which may promote better integration of the implant into bone tissue.

Bioactivity of the coating: Spectral analysis of the coating revealed the presence of characteristic bands for the functional groups Ti-O-Ti, Si-O, C-H, C=C, Si-O-Ti, and PO_4^{3-}. The presence of these chemical groups suggests that the coating exhibits bioactive properties, which is particularly important for medical applications. The hydroxyapatite in the coating may promote the osteointegration process, while the silane components ensure its stability.

Improved corrosion resistance: The application of the VTMS-HAp-VTMS-VTMS coating on the titanium surface significantly improved the corrosion resistance of the tested material. These effects were evident in the shift of the corrosion potential toward more positive values, the reduction in cathodic and anodic current density, and the increase in polarization resistance, confirming the effectiveness of the applied technology in corrosion protection.

Innovation of the technology used: The use of sol–gel technology to produce VTMS/HAp/VTMS/VTMS coatings, based on vinyltrimethoxysilane (VTMS) and hydroxyapatite (HAp), represents an innovative approach in biomaterials engineering, opening up new possibilities for the production of implant materials. This method, which has not been previously described in the literature, demonstrates high efficiency and offers significant potential for further research and applications.

5. Patents

Patent application no. P.440129.

Author Contributions: Conceptualization, A.D.; methodology, A.D.; formal analysis A.D. and O.K.; investigation, A.D. and O.K.; writing—original draft preparation, A.D. and O.K.; writing—review and editing, A.D. and O.K.; visualization, A.D. and O.K.; supervision, A.D. All authors have read and agreed to the published version of the manuscript.

Funding: This research received no external funding.

Institutional Review Board Statement: Not applicable.

Informed Consent Statement: Not applicable.

Data Availability Statement: The original contributions presented in this study are included in the article. Further inquiries can be directed to the corresponding authors.

Conflicts of Interest: The authors declare no conflicts of interest.

References

1. Chouirfa, H.; Bouloussa, H.; Migonney, V.; Falentin-daudré, C. Review of titanium surface modification techniques and coatings for antibacterial applications. *Acta Biomater.* **2019**, *83*, 37–54. [CrossRef]
2. Long, M.; Rack, H.J. Titanium alloys in total joint replacement—A materials science perspective. *Biomaterials* **1998**, *19*, 1621–1639. [CrossRef]
3. Niinomi, M.; Kuroda, D.; Fukunaga, K.; Morinaga, M.; Kato, Y.; Yashiro, T.; Suzuki, A. Corrosion wear fracture of new β type biomedical titanium alloys. *Mater. Sci. Eng. A* **1999**, *263*, 193–199. [CrossRef]
4. Niinomi, M. Recent metallic materials for biomedical applications. *Metall. Mater. Trans. A* **2002**, *33*, 477–486. [CrossRef]
5. Elias, C.N.; Lima, J.H.C.; Valiev, R.; Meyers, M.A. Biomedical applications of titanium and its alloys. *J. Miner. Met. Mater. Soc.* **2008**, *60*, 46–49. [CrossRef]
6. Surowska, B.; Bieniaś, J. Composite layers on titanium and Ti6Al4V alloy for medical applications. *J. Achiev. Mater. Manuf. Eng.* **2010**, *43*, 162–169.
7. Bai, Y.; Li, S.J.; Prima, F.; Hao, Y.L.; Yang, R. Electrochemical corrosion behavior of Ti-24Nb-4Zr-8Sn alloy in a simulated physiological environment. *Appl. Surf. Sci.* **2012**, *258*, 4035–4040. [CrossRef]
8. De Assis, S.L.; Wolynec, S.; Costa, I. Corrosion characterization of titanium alloys by electrochemical techniques. *Electrochim. Acta* **2006**, *51*, 1815–1819. [CrossRef]
9. Browne, M.; Gregson, P.J. Effect of mechanical surface pretreatment on metal ion release. *Biomaterials* **2000**, *21*, 385–392. [CrossRef]
10. Hübler, R.; Cozza, A.; Marcondes, T.L.; Souza, R.B.; Fiori, F.F. Wear and corrosion protection of 316-L femoral implants by deposition of thin films. *Surf. Coat. Technol.* **2001**, *142–144*, 1078–1083. [CrossRef]
11. Günzel, R.; Mändl, S.; Richter, E.; Liu, A.; Tang, B.Y.; Chu, P.K. Corrosion protection of titanium by deposition of niobium thin films. *Surf. Coat. Technol.* **1999**, *116–119*, 1107–1110. [CrossRef]

12. Sella, C.; Martin, J.C.; Lecoeur, J.; Le Chanu, A.; Harmand, M.F.; Naji, A.; Davidas, J.P. Biocompatibility and corrosion resistance in biological media of hard ceramic coatings sputter deposited on metal implants. *Mater. Sci. Eng. A* **1991**, *139*, 49–57. [CrossRef]
13. Hübler, R. Characterisation of gradient interfaces in thin film multilayers used to protect orthopedic implants. *Surf. Coat. Technol.* **1999**, *116–119*, 1116–1122. [CrossRef]
14. Przybyszewska-Doroś, I.; Okrój, W.; Walkowiak, B. Surface modifications of metallic implants. *Inżynieria Biomateriałów* **2005**, *43–44*, 52–62.
15. Liu, X.; Chu, P.K.; Ding, C. Surface modification of titanium, titanium alloys, and related materials for biomedical applications. *Mater. Sci. Eng. R Rep.* **2004**, *47*, 49–121. [CrossRef]
16. Catauro, M.; Papale, F.; Bollino, F. Characterization and biological properties of TiO_2/PCL hybrid layers prepared via sol-gel dip coating for surface modification of titanium implants. *J. Non-Cryst. Solids* **2015**, *415*, 9–15. [CrossRef]
17. Javadi, A.; Solouk, A.; Nazarpak, M.H.; Bagheri, F. Surface engineering of titanium-based implants using electrospraying and dip coating methods. *Mater. Sci. Eng. C* **2019**, *99*, 620–630. [CrossRef]
18. He, G.; Hu, J.; Wei, S.C.; Li, J.H.; Liang, X.H.; Luo, E. Surface modification of titanium by nano-TiO_2/HA bioceramic coating. *Appl. Surf. Sci.* **2008**, *255*, 442–445. [CrossRef]
19. Dudek, A.; Adamczyk, L. Properties of hydroxyapatite layers used for implant coatings. *Opt. Appl.* **2013**, *43*, 143–151.
20. Sasikumar, Y.; Indira, K.; Rajendran, N. Surface Modification Methods for Titanium and Its Alloys and Their Corrosion Behavior in Biological Environment: A Review. *J. Bio- Tribo-Corros.* **2019**, *5*, 36. [CrossRef]
21. Li, J.; Zhou, P.; Attarilar, S.; Shi, H. Innovative Surface Modification Procedures to Achieve Micro/Nano-Graded Ti-Based Biomedical Alloys and Implants. *Coatings* **2021**, *11*, 647. [CrossRef]
22. Asad, M.; Sana, M. Potential of titanium based alloys in the biomedical sector and their surface modification techniques: A review. *Proc. Inst. Mech. Eng. Part C J. Mech. Eng. Sci.* **2023**, *237*, 5503–5532. [CrossRef]
23. Han, X.; Ma, J.; Tian, A.; Wang, Y.; Li, Y.; Dong, B.; Tong, X.; Ma, X. Surface modification techniques of titanium and titanium alloys for biomedical orthopaedics applications: A review. *Colloids Surf. B Biointerfaces* **2023**, *227*, 113339. [CrossRef]
24. Kurup, A.; Dhatrak, P.; Khasnis, N. Surface modification techniques of titanium and titanium alloys for biomedical dental applications: A review. *Mater. Proc.* **2021**, *39*, 84–90. [CrossRef]
25. Xue, T.; Attarilar, S.; Liu, S.; Liu, J.; Song, X.; Li, L.; Zhao, B.; Tang, Y. Surface Modification Techniques of Titanium and its Alloys to Functionally Optimize Their Biomedical Properties: Thematic Review. *Front. Bioeng. Biotechnol.* **2020**, *8*, 603072. [CrossRef]
26. Zhang, L.-C.; Chen, L.-Y.; Wang, L. Surface Modification of Titanium and Titanium Alloys: Technologies, Developments, and Future Interests. *Adv. Eng. Mater.* **2019**, *22*, 1901258. [CrossRef]
27. Kwok, C.T.; Wong, P.K.; Cheng, F.T.; Man, H.C. Characterization and corrosion behavior of hydroxyapatite coatings on Ti6Al4V fabricated by electrophoretic deposition. *Appl. Surf. Sci.* **2009**, *255*, 6736–6744. [CrossRef]
28. Chou, B.-Y.; Chang, E. Influence of deposition temperature on mechanical properties of plasma-sprayed hydroxyapatite coating on titanium alloy with ZrO_2 intermediate layer. *J. Therm. Spray Technol.* **2003**, *12*, 199–207. [CrossRef]
29. Family, R.; Solati-Hashjin, M.; Nik, S.N.; Nemati, A. Surface modification for titanium implants by hydroxyapatite nanocomposite. *Casp. J. Intern. Med.* **2012**, *3*, 460–465.
30. Kierat, O.; Dudek, A.; Adamczyk, L. Dependence of protective properties of silane coatings made on titanium Grade 2 and Ti6Al4V substrate on the component—Vinyltrimethoxysilane (VTMS). *Corros. Prot.* **2020**, *63*, 364–368.
31. Kierat, O.; Dudek, A.; Adamczyk, L. The Effect of the Corrosion Medium on Silane Coatings Deposited on Titanium Grade 2 and Titanium Alloy Ti13Nb13Zr. *Materials* **2021**, *14*, 6350. [CrossRef] [PubMed]
32. *ASTM D3359*; Standard Test Methods for Measuring Adhesion by Tape Test. ASTM INTERNATIONAL: West Conshohocken, PA, USA, 2017. [CrossRef]
33. Srinivasan, A.; Rajendran, N. Surface characteristics, corrosion resistance and MG63 osteoblast-like cells attachment behavior of nano SiO_2-ZrO_2 coated 316L stainless steel. *RSC Adv.* **2015**, *5*, 26007–26016. [CrossRef]
34. Nguyen, V.G.; Thai, H.; Mai, D.H.; Tran, H.T.; Tran, D.L.; Vu, M.T. Effect of titanium dioxide on the properties of polyethylene/TiO_2 nanocomposites. *Compos. Part B Eng.* **2013**, *45*, 1192–1198. [CrossRef]
35. Iribarren-Mateos, J.I.; Buj-Corral, I.; Vivancos-Calvet, J.; Alemán, C.; Iribarren, J.I.; Armelin, E. Silane and epoxy coatings: A bilayer system to protect AA2024 alloy. *Prog. Org. Coat.* **2015**, *81*, 47–57. [CrossRef]
36. Qin, Z.; Li, D.; Zhang, W.; Yang, R. Surface modification of ammonium polyphosphate with vinyltrimethoxysilane: Preparation, characterization, and its flame retardancy in polypropylene. *Polym. Degrad. Stab.* **2015**, *119*, 139–150. [CrossRef]
37. Hu, J.-M.; Liu, L.; Zhang, J.-Q.; Cao, C.-N. Electrodeposition of silane films on aluminum alloys for corrosion protection. *Prog. Org. Coat.* **2007**, *58*, 265–271. [CrossRef]
38. Romero-Serrano, M.; Romero-Ruiz, M.-M.; Herrero-Climent, M.; Rios-Carrasco, B.; Gil-Mur, J. Correlation between Implant Surface Roughness and Implant Stability: A Systematic Review. *Dent. J.* **2024**, *12*, 276. [CrossRef] [PubMed]
39. Almas, K.; Smith, S.; Kutkut, A. What is the best micro and macro dental implant topography? *Dent. Clin.* **2019**, *63*, 447–460. [CrossRef]
40. Lyon, S.; Richardson, T.; Cottis, B.; Lindsay, R.; Scantlebury, D.; Stott, H.; Graham, M. *Shreir's Corrosion*, 4th ed.; Elsevier BV: Amsterdam, The Netherlands, 2010; p. 3450.
41. Edraki, M.; Zaarei, D.; Sabeeh Hasan, I. The Impact of Green Corrosion Inhibitors on the Protection Performance of Hybrid Silane Sol-Gel Coatings: A Review. *Chem. Rev. Lett.* **2023**, *6*, 428–441.

42. Asadi, N.; Naderi, R. Nanoparticles incorporated in silane sol-gel coatings. In *Corrosion Protection at the Nanoscale*; Rajendran, S., ANH Nguyen, T., Yeganeh, M., Li, Y., Eds.; Elsevier: Amsterdam, The Netherlands, 2020; Chapter 23; pp. 451–471.
43. Khezri, A.; Edjlali, L.; Es'haghi, M.; Taghi-Bardini, M.; Basharnavaz, H. Addition of Schiff Bases to Hybrid Silane Sol-Gel Coatings: An Efficient Strategy to Develop an Active System for Corrosion Protection of Copper. *J. Mater. Eng. Perform.* **2023**, *32*, 10740–10749. [CrossRef]
44. Ghodrati, M.; Mousavi-Kamazani, M.; Bahrami, Z. Synthesis of superhydrophobic coatings based on silica nanostructure modified with organosilane compounds by sol-gel method for glass surfaces. *Sci. Rep.* **2023**, *13*, 548. [CrossRef] [PubMed]

materials

MDPI

Article

Synthesis of Amphiphilic Polyacrylates as Peelable Coatings for Optical Surface Cleaning

Daofeng Zhu [1], Hao Huang [1], Anqi Liang [1], Yanling Yang [1], Baohan He [1], Abbas Ahmed [2], Xiaoyan Li [1,*], Fuchuan Ding [1,*] and Luyi Sun [2,*]

[1] College of Chemistry and Materials Science & Fujian Key Laboratory of Polymer Science, Fujian Normal University, Fuzhou 350007, China; qsz20211446@student.fjnu.edu.cn (D.Z.); qsx20210995@student.fjnu.edu.cn (H.H.); qsx20220946@student.fjnu.edu.cn (A.L.); 107062020010@student.fjnu.edu.cn (Y.Y.); 107062020024@student.fjnu.edu.cn (B.H.)

[2] Polymer Program, Institute of Materials Science and Department of Chemical & Biomolecular Engineering, University of Connecticut, Storrs, CT 06269, USA; abbas.ahmed@uconn.edu

* Correspondence: lixy76@fjnu.edu.cn (X.L.); dingfc@fjnu.edu.cn (F.D.); luyi.sun@uconn.edu (L.S.); Tel.: +86-139-0691-6919 (X.L.); +86-177-0591-1106 (F.D.); +1-860-486-6895 (L.S.)

Abstract: Optical instruments require extremely high precision, and even minor surface contamination can severely impact their performance. Peelable coatings offer an effective and non-damaging method for removing contaminants from optical surfaces. In this study, an amphiphilic polyacrylate copolymer (PMLEA) was synthesized via solution radical copolymerization using the lipophilic monomer lauryl acrylate (LA) and hydrophilic monomers ER-10, methyl methacrylate (MMA), and butyl acrylate (BA). The structure and molecular weight of the copolymer were characterized using Fourier transform infrared spectroscopy (FTIR), nuclear magnetic resonance (NMR), and gel permeation chromatography (GPC). The hydrophilic–lipophilic balance, surface tension, and wettability of the copolymer were analyzed through water titration, the platinum plate method, and liquid contact angle tests. The cleaning performance of the copolymer coating on quartz glass surface contaminants was evaluated using optical microscopy and Ultraviolet-Visible Near-Infrared (UV-Vis-NIR) spectroscopy. The study examined the effect of varying the ratio of LA to ER-10 on the hydrophilicity, lipophilicity, cleaning efficiency, and mechanical properties of the copolymer coating. The results showed that when the mass ratio of LA to ER-10 was 1:2, the synthesized copolymer exhibited optimal performance in removing dust, grease, and fingerprints from quartz glass surfaces. The coating had a tensile strength of 2.57 MPa, an elongation at break of 183%, and a peeling force of 2.07 N m^{-1}.

Keywords: optical surface cleaning; amphiphilic polymer; peelable coating

1. Introduction

The cleanliness of optical components' surfaces is paramount. Contaminants can potentially absorb or scatter light, leading to the significant degradation of optical performance [1], including a decrease in transmittance [2], a reduction in laser-induced damage threshold [3,4], and serving as the initiators of laser-induced damage [5,6]. Dust, grease, and fingerprints represent the primary sources of contamination for optical devices [7,8]. Currently, dry-cleaning and wet-cleaning systems stand as the predominant methods for cleaning optical components [9,10]. Dry-cleaning techniques encompass methods such as dry ice blasting [11–13], mechanical wiping [14], and laser cleaning [8,15,16], while wet-cleaning includes steam cleaning [17] and solvent immersion cleaning [18]. However, dry-cleaning necessitates wiping the surface of the object, which poses a risk of scratching if there are particles of contaminants present. Conversely, after wet-cleaning, a residual smear forms a surface layer that can adversely affect the focus and contrast of optical elements [19,20]. In contrast, peelable coating cleaning technology offers several advantages, including high efficiency without secondary pollution, the avoidance of surface damage, and environmental friendliness [21].

Peelable coating serves as a temporary protective layer that can be easily removed from the substrate surface either at the end of its lifespan or after fulfilling its intended purpose [22]. During the removal process, it is crucial that the cohesive force within the strippable film exceeds the adhesive forces between the film and the substrate. This ensures that the film can be peeled off completely, leaving behind no residue [23,24]. Peeling off the functional coating and damaging the device's surface would occur if the peeling force of the cleaning coating was greater than the functional coating's adhesive force on the surface of the optical element [25].

Polymers commonly used in strippable coatings include vinyl polymers [26,27], polyurethanes [28,29], acrylics [30,31], and cellulose and its derivatives [32,33]. Choosing the right film-forming polymer and solvent for peelable coatings on optical surfaces is critical. Several conditions should be met by peelable cleaning coatings on optical surfaces [34]: rapid solvent-drying and short polymer-curing time, adaptability to different substrates, sufficient mechanical properties, the ease of application, quick and complete stripping, and protection of object surfaces while meeting cleaning requirements.

The effectiveness of cleaning depends on several factors, including temperature, hydrodynamic force, formulation, and cleaning time [35]. Contaminants on optical surfaces are often either hydrophilic or lipophilic. The amphiphilic nature of polymers facilitates the adhesion and encapsulation of contaminants, which increases the cleaning effectiveness [36, 37].

Acrylate-based coatings are preferred for their durability, hardness, and scratch resistance, while the incorporation of copolymer monomers enhances functional versatility and adjustability compared to alternative coatings [38,39].

In this study, amphiphilic polyacrylates intended as film-forming materials were synthesized via solution polymerization. The primary monomers, methyl methacrylate and butyl acrylate, were combined with lauryl acrylate and ER-10 as functional monomers. Methyl methacrylate contributes strength, hardness, and adhesion to the coating, while butyl acrylate imparts flexibility, resulting in a balance of softness and hardness that yields excellent mechanical performance. Lauryl acrylate's lengthy carbon chain endues the coating with lipophilicity and low adhesive force, which ensure that the cleaning coating can be easily removed without damaging the functional coating of the optical device, while the ER-10 surfactant adds hydroxyl groups to the resultant polymer to increase its degree of hydrophilicity. This coating improves the wetting, adhesion, and encapsulation of both aqueous and oily contaminants by enhancing the amphiphilicity of the copolymer. This allows it to effectively penetrate the interface between the contaminants and the material surface during cleaning, lifting the contaminants and incorporating them into the coating. When the coating is peeled away, the contaminants are removed along with it. The successful synthesis of the product was confirmed through the characterizations using Fourier transform infrared spectroscopy (FTIR), nuclear magnetic resonance (NMR), and gel permeation chromatography (GPC) techniques. The resulting coating exhibits a tensile strength of 3.47 MPa and a peel strength of less than 3.20 N m^{-1}. These polyacrylate coatings demonstrate excellent performance in removing contaminants from quartz glass surfaces, leaving no visible residue under an optical microscope after cleaning. Furthermore, the transmittance of the cleaned quartz is restored to a level close to that of the original clean quartz.

2. Experiment

2.1. Materials

Methyl methacrylate (MMA), butyl acrylate (BA), 2,2′-azobis(2-methylpropionitrile) (AIBN), lauryl acrylate (LA), petroleum ether, propylene glycol methyl ether acetate (PMA), Span 80, Tween 20, ethanol, 1,4-dioxane, toluene, 1,3-dioxolane, and butanone were sourced from Sinopharm Chemical Reagent Co., Ltd. (Shanghai, China). The above reagents are analytically pure except for Span 80 and Tween 20, which are chemically pure. The SiO_2 particles (AEROSIL® R972) were from Evonik Industries AG (Essen, Germany). The silicone grease was from Sinopec Lubricants Co., Ltd. (Beijing, China). The ER-10 emulsifier was procured from Nippon Adeka Co., Ltd. (Tokyo, Japan). The specific structure of ER-10 is shown in Figure S2 (Support Information), where R is an alkyl group.

2.2. Preparation of Polyacrylates

The polyacrylates were synthesized via solution polymerization using MMA, BA, LA, and ER-10 as monomers. The synthetic formulations are detailed in Table 1. The synthesis of polyacrylates is outlined in Figure 1, while the preparation details are provided in Supplementary Figure S1. The monomers, along with 38.40 g of PMA as the solvent, were combined in a 100.0 mL flask. The reaction was conducted in a water bath at 70 °C for 5 h with AIBN as the initiator. After the reaction, the polyacrylate sample was purified by first mixing the reacted solution into a 50% (v/v) water–ethanol solution to precipitate white flocculent. Subsequently, the product underwent washing with petroleum ether three times, followed by soaking in methanol for 24 h to obtain the purified polyacrylate sample. Finally, the polyacrylate after vacuum drying at 50 °C was dissolved in a 1:1 mass ratio of 1,3-dioxolane/butanone mixture to achieve a solution of 10 wt%.

Table 1. Formulations to synthesize acrylate polymers.

Sample	LA (g)	ER-10 (g)	MMA (g)	BA (g)
PMLA	7.00	0		
PMLEA-1	3.50	3.50		
PMLEA-2	2.34	4.66		
PMLEA-3	1.75	5.25	12.00	8.00
PMLEA-4	1.40	5.60		
PMLEA-5	1.17	5.83		

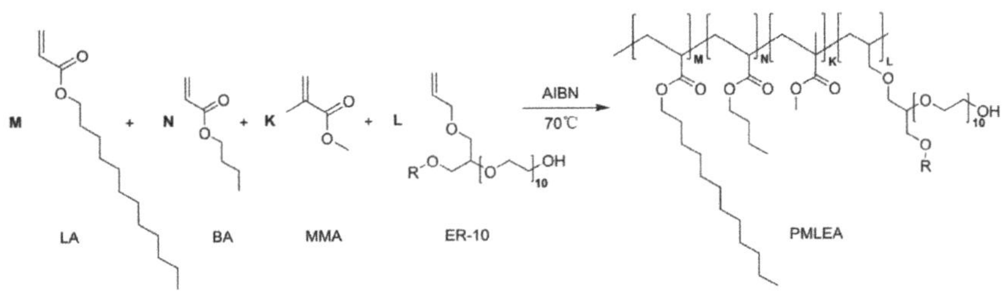

Figure 1. Synthesis reaction of PMLEA.

2.3. Characterization and Performance Evaluation

2.3.1. FTIR and NMR Characterization

The Fourier transform infrared (FTIR) spectra of approximately 5.0 µm thick polyacrylate films coated on single-crystal silicon wafers were collected using a Nicolet iS50 spectrophotometer (Thermo Fisher Scientific, Waltham, MA, USA) with a wavelength scanning range of 4000–400 cm^{-1}. The NMR spectra of the polyacrylates were obtained using a Bruker 400 MHz spectrometer (Billerica, MA, USA), employing $CDCl_3$ as the solvent.

2.3.2. Gel Permeation Chromatography (GPC)

The gel permeation chromatography of the copolymer samples was performed using an Agilent 1260 (Santa Clara, CA, USA) liquid chromatograph equipped with a refractive index (RI) detector. The column setup included Agilent PLgel 5 µm Mixed-c, 10 µm Mixed-b, and PLgel 5 µm Guard in series. THF was used as the mobile phase with a flow rate set at 1 mL min^{-1}.

2.3.3. Differential Scanning Calorimetry (DSC)

The glass transition temperatures of the polyacrylates were determined using DSC (METTLER DSC3, Greifensee, Switzerland) in a temperature range of −20 to 200 °C at a rate of 10 °C/min under a N_2 atmosphere.

2.3.4. Hydrophilic–Lipophilic Balance (HLB) of the Polyacrylates

HLB, initially proposed by Griffin [40,41], represents an equilibrium relationship between the hydrophilic and lipophilic groups in surfactants concerning their size and strength. In this study, the HLBs of the polyacrylates were determined using the aqueous titration method developed by Greenwald [42]. The specific test procedures are as follows: A total of 1.00 g of a mixture of Span 80 and Tween 20 at various mass ratios was charged into a 100 mL conical flask, and then 30.0 mL of a mixture of 1,4-dioxane and toluene with a volume ratio of 96:4 was added to the conical flask. The solution was titrated with deionized water by shaking the conical flask until the solution became cloudy at 30 °C. The volume of titrating water was recorded as W_1. Then, the relationship between log(W_1) and HLB was plotted using a polynomial equation. The water volume (W_2) required for the polymer solution of the 1,4-dioxane and toluene mixture was determined in the same way. Finally, the HLB of each polyacrylate was calculated by substituting W_2 into the fitting Equation (1).

$$Y = A_1 * x^3 + A_2 * x^2 + A_3 * x + Intercept \quad (1)$$

2.3.5. Surface Tension of Polymer Solutions

The surface tension of each polyacrylate solution was measured as the stable test values recorded by an automatic surface tension meter (K20, Kruss, Hamburg, Germany) using a platinum plate (24.0 mm × 10.0 mm × 0.1 mm) in contact with the surface of 10 wt% polyacrylate solution in a 1:1 mass ratio of 1,3-dioxolane/butanone mixture.

2.3.6. Polyacrylate Wettability and Amphiphilicity

The wettability of each polyacrylate solution was characterized by recording the contact angle after 3 s for a drop of the polymer solution placed on the quartz glass surface using a syringe with a contact angle tester (DSA25E, Kruss, GER).

Additionally, the amphiphilicity of each polyacrylate was assessed by measuring the contact angles of 1.0 μL drops of deionized water and diiodomethane on the surface of the polyacrylate coating using a contact angle tester (DSA25E, Kruss, GER).

2.3.7. Tensile Properties of the Polyacrylate Coatings

The tensile strength of the polyacrylates, which were sized at 70.0 mm × 25.0 mm × 0.1 mm, was tested using a universal testing machine (LR5K Plus, LLOYD, Bognor Regis, UK) equipped with a 100 N load cell. The test was conducted at a rate of 50.0 mm/min.

2.3.8. The 180° Peel Test of Polyacrylate Coatings

The 180° peel test of the polyacrylate coatings involved peeling off coatings sized at 50.0 mm × 25.0 mm × 0.05 mm at a 180° angle, with a speed of 10.0 mm/min, using a universal testing machine (LR5K Plus, LLOYD, UK) equipped with a 100 N load cell.

2.3.9. Coating Cleaning Effect Evaluation

In this study, three types of contamination: SiO_2 particles, silicone grease, and fingerprints were targeted for cleaning on quartz glass using a peelable coating. First, silicone grease was dissolved in toluene, while SiO_2 particles (15.0 nm in diameter) were dispersed in anhydrous ethanol. These dispersions were then used to contaminate clean quartz glass, along with fingerprints, to create contaminated quartz glass. Next, a polymer solution was applied to the contaminated glass, allowed to dry completely, and then peeled off to remove the contaminants. Finally, cleaning comparison photographs were obtained using an optical microscope (EVDM10103-M32, Easy View, Shenzhen, Guangdong, China), and the transmission spectra of the quartz glass before and after cleaning were obtained using a UV-Vis-NIR spectrophotometer (LAMBDA750S, PerkinElmer, Waltham, MA, USA).

3. Results and Discussion

Figure 2a displays the FTIR spectra of the polyacrylate samples PMLA and PMLEA-2. The peaks observed at 2953 and 1358 cm^{-1} correspond to the stretching and deformation vibrations of methyl (-CH_3) groups, respectively. The stretching vibration and deformation vibration of the methylene (-CH_2) group are present at 2877 and 1450 cm^{-1}, respectively. The characteristic peaks of the stretching vibration of carbonyl (C=O) and ether (C-O-C) groups are observed at 1731 and 1148 cm^{-1}, respectively. The pronounced enhancement of the peak at 1148 cm^{-1} in sample PMLEA-2 is attributed to the presence of polyethylene oxide groups in ER-10.

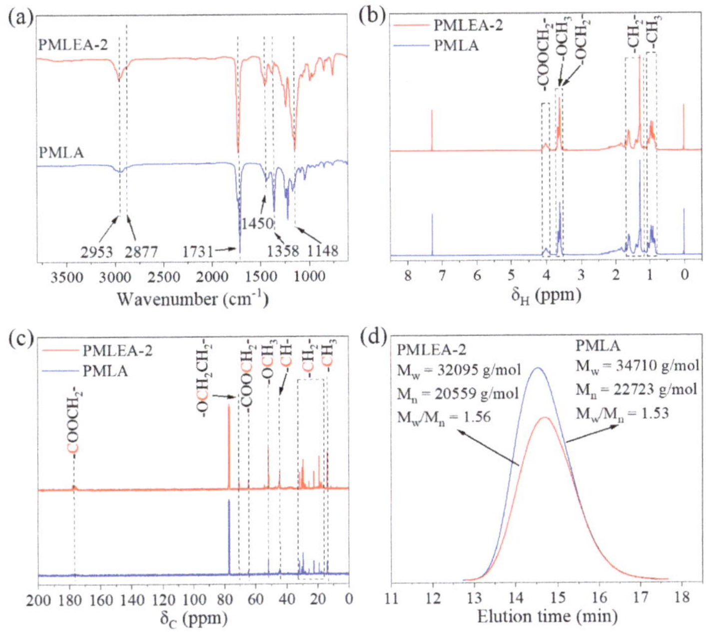

Figure 2. (**a**) FTIR spectra, (**b**) ^{1}H-NMR spectra, (**c**) ^{13}C-NMR spectra, and (**d**) GPC chromatograms of polyacrylates PMLA and PMLEA-2.

Figure 2b,c depict the ^{1}H-NMR and ^{13}C-NMR spectra of the polyacrylate samples PMLA and PMLEA-2, respectively. The peaks at δ_H = 3.61 ppm and δ_C = 51.73 ppm are the corresponding hydrogen and carbon shift peaks of the -OCH_3 groups in the polyacrylates, and the peaks at δ_H = 4.01 ppm and δ_C = 64.50 ppm are the chemical shifts of hydrogen and carbon of the -CH_2- adjacent to the lipid group in the polyacrylates. The peak at δ_C = 176.73 ppm is the carbon shift peak of the lipid group -COO- in the copolymer. Additionally, the ^{13}C-NMR spectrum of PMLEA-2 exhibits a chemical shift of 70.48 ppm for vinyl oxide. Meanwhile, the proportion of each copolymer component was estimated from the characteristic peak areas in the ^{1}H NMR spectra (Figure S3) of the polyacrylate samples. The results are shown in Table S1. It can be seen that the ER-10 content in the polyacrylates is directly proportional to the formulated amount. Notably, the percentage of ER-10 in the synthesized polyacrylates is significantly lower than in the mixture of the monomers. This is due to ER-10's significantly lower reactivity in comparison to other monomers.

As depicted in Figure 2d, which shows the GPC chromatograms of the polyacrylate samples PMLA and PMLEA-2 (please see more details in Figure S4). The number average molecular weight of the polyacrylate samples exceeds 2.0×10^5 g/mol. Furthermore, the occurrence of chain development during the copolymerization of the four monomers in a one-pot synthesis resulted in a diverse distribution of polymer chains. Despite this, a molecular weight distribution below 1.6 and a single peak-shaped molecular weight distribution were observed in each polyacrylate sample, which indicates that these four monomers have good copolymerization ability, suggesting that the polyacrylates should exhibit the desired tensile and viscoelastic properties for use as a coating.

Figure 3 presents the DSC thermograms of the polyacrylate samples PLMA and PMLEA-2, which show that the glass transition temperatures of PMLA and PMLEA-2 are 11.3 and 29.2 °C, respectively. The increase in the glass transition temperature indicates the increase in the rigidity of the polymer chain segments and the consequent increase in the strength of the polyacrylate coating. This is due to the high molecular weight of ER-10 monomers. As a result, the flexibility of the polymer chain of PMLEA-2 is reduced. In addition, the hydroxyl groups in ER-10 readily form hydrogen bonds, which greatly increase the rigidity of the polymer chains. The glass transition temperature of the sample PLMEA-2 is higher than room temperature, which is ideal for the coating's practical use.

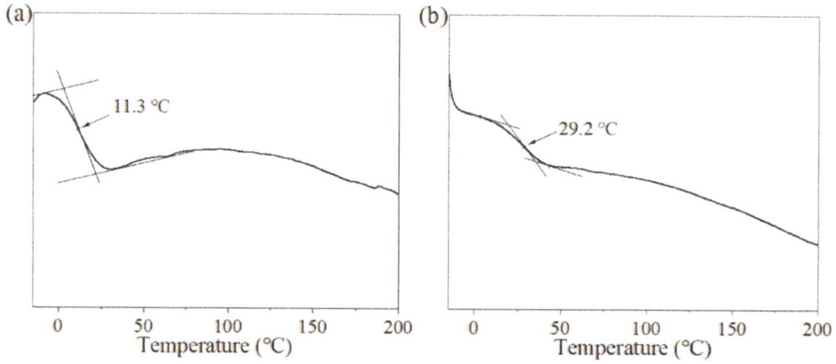

Figure 3. DSC thermograms of polyacrylates: (**a**) PMLA and (**b**) PMLEA-2.

The relationship between the $\log(W_1)$ and HLB of Span 80 and Tween 20 in a mixed solution of 1,4-dioxane/toluene is depicted in Figure 4, with the red curve representing the fitted equation. The curve equation is as follows:

$$Y = 76.52x^3 - 254.36x^2 + 289.00x - 95.228 \quad (2)$$

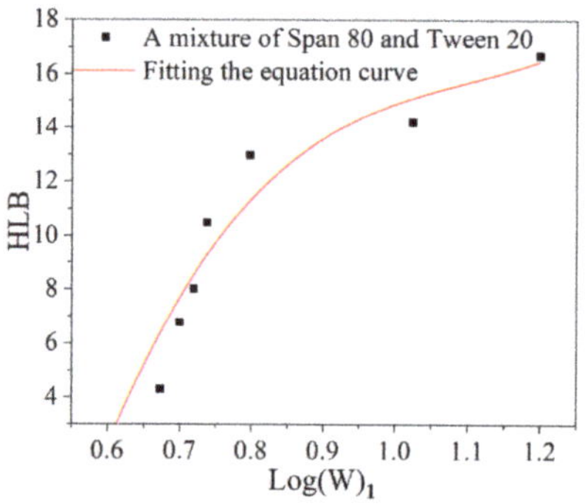

Figure 4. $\text{Log}(W_1)$ versus HLB of Span 80 and Tween 20 in a mixed solution of 1,4-dioxane/toluene.

The corresponding HLB values are obtained by substituting the water volume value in each polyacrylate sample in Equation (2) and are listed in Table 2. The ratio of hydrophilic to lipophilic groups in the polymer determines the magnitude of the HLB values, and a larger HLB value denotes a more lipophobic and hydrophilic polymer. Therefore, to design an amphiphilic polymer with the effect of reducing surface tension, the HLB value of the polymer should be maintained within a reasonable range. In the sample PMLA, there is no presence of hydrophilic groups, resulting in an HLB value of 0 [41]. Moreover, the hydrophilicity of the polyacrylates increases as the percentage of ER-10 in the samples increases, while simultaneously, the lipophilicity of the polyacrylate samples decreases. An HLB value between 7 and 9 indicates that the polyacrylate sample possesses excellent infiltration capacity, facilitating the effective penetration and wetting of stains for optimal cleaning results [43].

Table 2. Volumes of water in the polymer samples and the corresponding HLB values.

Sample	W_2 (mL)	Log(W_2)	HLB
PMLA	3.6	0.56	0
PMLEA-1	4.2	0.62	4.61
PMLEA-2	4.8	0.68	7.78
PMLEA-3	5.2	0.71	9.39
PMLEA-4	5.4	0.73	10.1
PMLEA-5	5.8	0.76	11.2

Figure 5a presents the surface tension results of the polymer solutions. The addition of a polyacrylate resulted in a decrease in the solution's surface tension. Moreover, an increased concentration of ER-10 in the polyacrylate contributed to a reduction in the surface tension of the resultant polymer solution, enhancing its wettability. Consequently, the polymer solution spreads more easily on the substrate surface. Figure 5b presents the contact angles of the polymer solutions on the surface of quartz glass. A larger contact angle indicates the lower surface wettability of a polymer solution. This reduces the contact area between the contaminant and the coating, leading to less effective contaminant removal. However, the contact angles of the six polymer solutions on the quartz glass surface are all less than 40°, indicating that all of them can effectively wet the surface of a clean quartz glass. A polyacrylate with higher hydrophilicity possesses better wettability on a clean quartz glass surface, making it easier to spread on a quartz glass surface. This is consistent with the results of the surface tension of the polymer solutions.

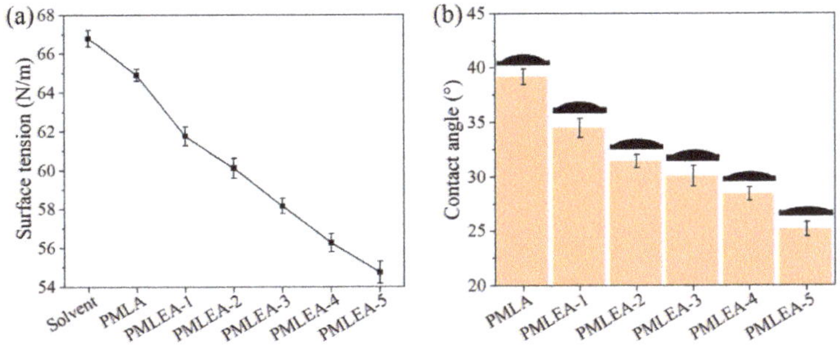

Figure 5. (**a**) Surface tension values of the polymer solutions; (**b**) contact angles of the polymer solution on quartz glasses.

The lower the contact angle between water or oil on its surface, the more hydrophilic or lipophilic the coating is. This promotes easier wetting of stains, allowing contaminants to adhere more effectively to the coating or be partially encapsulated by the coating. In the contact angle test depicted in Figure 6, due to the predominance of lipophilic monomers in the resultant polyacrylates, the coatings exhibit significant lipophilicity. They can effectively impregnate and coat oily stains. However, as the amount of ER-10 increases and the amount of LA decreases, the hydrophilicity of the polyacrylate samples increases while their lipophilicity decreases. This indicates a reduction in the polyacrylate's ability to adhere to oily stains, while its ability for aqueous stains increases. Furthermore, the water contact angles of the coatings are greater than 65°, providing temporary protection to the substrate from secondary contamination by airborne moisture.

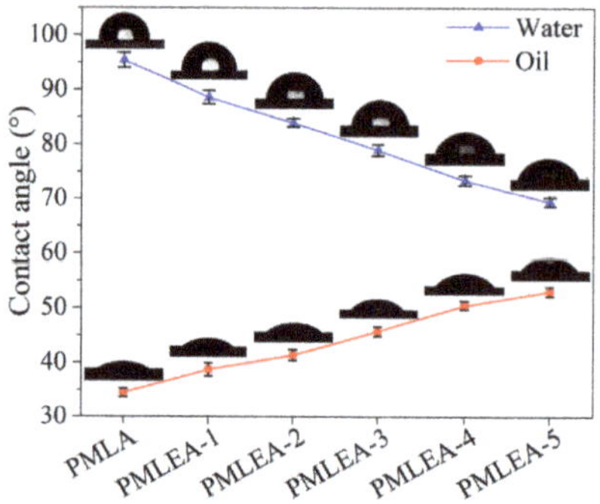

Figure 6. Contact angles of the polyacrylate coatings.

Table 3 provides the mechanical characterization results of the polyacrylate coatings. The PMLA coating is softer, with a tensile strength of just 1.35 MPa and an elongation at break of 275%. In comparison, the PLMEA-5 coating is brittle, with a tensile strength of 3.47 MPa and an elongation at break of 38.3%. This means that substantial variations in the toughness and strength of the coatings were observed in relation to the amount of LA and ER-10 in the polyacrylate samples. LA contributes to coating flexibility, with coating toughness increasing as the amount of LA in the polyacrylate sample rises. However, in the PMLEA series of samples, the flexibility of the polymer molecular chains is diminished due to the high molecular weight of ER-10 monomers. Additionally, the hydroxyl groups in ER-10 readily form hydrogen bonds, significantly enhancing the rigidity of the polymer's molecular chains. Consequently, coating toughness decreases and strength increases as the ER-10 content in the polyacrylate samples increases. Overall, the polyacrylate samples PMLEA-2 and PMLEA-3 exhibit balanced strength and toughness.

Table 3. Mechanical properties of polyacrylate coatings.

Sample	Stress (MPa)	Strain (%)	Peel Strength (N m^{-1})
PMLA	1.35 ± 0.07	275 ± 9	1.28 ± 0.11
PMLEA-1	1.95 ± 0.12	258 ± 7	1.78 ± 0.08
PMLEA-2	2.57 ± 0.04	183 ± 5	2.07 ± 0.11
PMLEA-3	2.87 ± 0.11	110 ± 2	2.46 ± 0.08
PMLEA-4	3.23 ± 0.07	76.6 ± 4.1	2.94 ± 0.13
PMLEA-5	3.47 ± 0.09	38.3 ± 3.7	3.20 ± 0.09

The PMLA coating's peel strength during coating peeling is just 1.28 N m^{-1}. However, an increase in the hydroxyl content of the polyacrylate samples results in the enhanced adhesion of the coatings to the quartz glass surface, leading to an increase in peel strength. As a result, the coating samples PMLEA-5 and PMLEA-4, which exhibited brittleness, tended to break during 180° peeling and could not be completely removed.

Figure 7 presents digital photographs showcasing the representative cleaning effect of PMLEA-2 on three contaminants on quartz glasses. From a macroscopic perspective, the contaminants on the surface of the quartz glass are effectively removed, with clearly defined boundaries between the uncleaned area on the left and the cleaned area on the right. Microphotographs of the boundary areas were utilized to compare the cleanliness of the different samples, as described below.

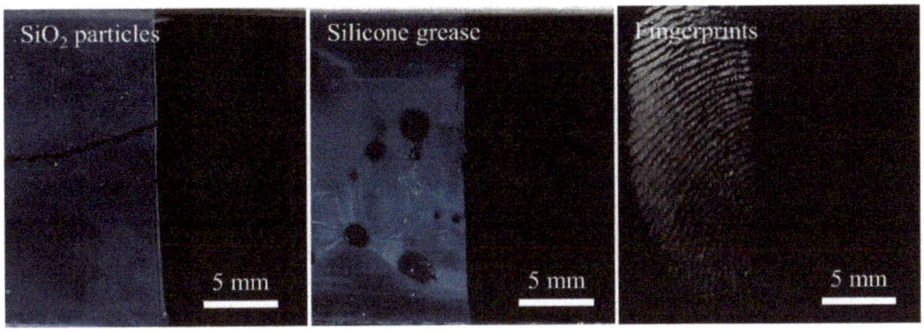

Figure 7. Representative digital photos of quartz glass before and after the removal of surface contamination by PMLEA-2.

Figure 8a displays the micrographs of the different polyacrylate samples for cleaning SiO_2 particles, demonstrating that each sample could effectively remove SiO_2 particles. In Figure 8b, the UV-Vis-NIR transmission spectra of the quartz glass before and after the cleaning are presented. The transmittance of the quartz glass is reduced after the contamination. However, the transmittance of the cleaned quartz glass can be restored to a level close to that of the initial clean quartz glass. This restoration in transmittance indicates the efficacy of the cleaning process. The low surface tension and good wetting properties of the polyacrylate solutions facilitate the adhesion and encapsulation of the SiO_2 particles, resulting in a robust cleaning performance.

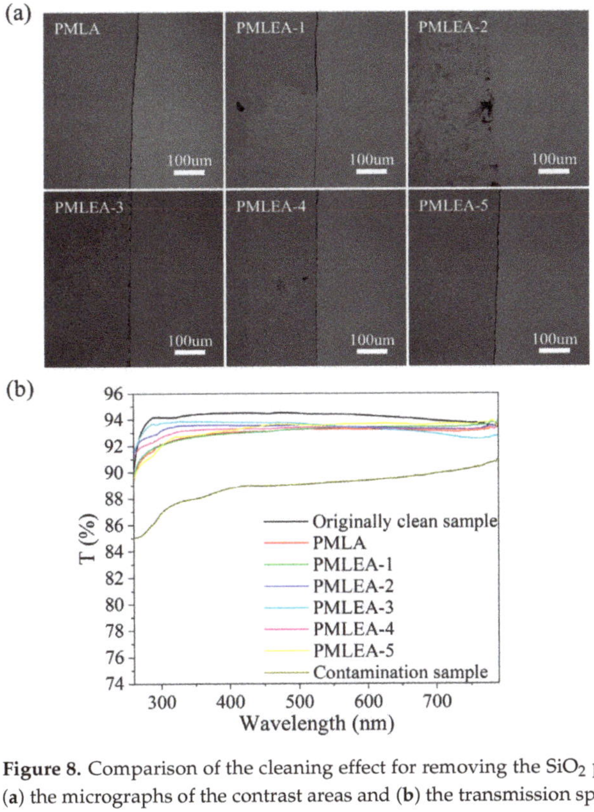

Figure 8. Comparison of the cleaning effect for removing the SiO_2 particles from the quartz glass: (**a**) the micrographs of the contrast areas and (**b**) the transmission spectra of the quartz glass before and after cleaning.

Figure 9a presents the micrographs illustrating the effect of silicone grease removal by the different polyacrylate samples, while Figure 9b displays the UV-Vis-NIR transmission spectra of the quartz glass before and after cleaning with the different polyacrylate samples. As the lipophilicity of the polyacrylate samples decreases, more contaminants remain on the surface of the quartz glass, resulting in decreased cleaning effectiveness. This phenomenon stems from the reduced lipophilicity of the polyacrylate samples, leading to a diminished ability of the polymer solutions to wet and penetrate oily contamination. Consequently, some of the polymer solutions are unable to fully cover and penetrate the surface of the contaminants and the contaminants nestled within, thereby compromising the cleaning efficacy of some of the polyacrylate samples.

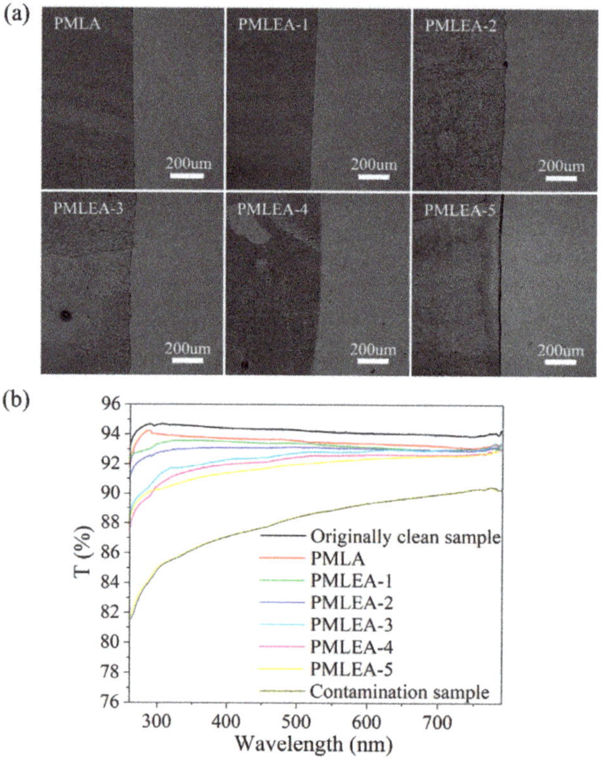

Figure 9. Comparison of the cleaning effect for removing silicone grease from the quartz glass: (**a**) micrographs of the contrast areas and (**b**) transmission spectra of the quartz glass before and after cleaning.

Figure 10a presents the micrographs illustrating the effect of the polyacrylate samples on the removal of fingerprints, while Figure 10b displays the transmission spectra of the quartz glass before and after cleaning with the different polyacrylate samples. The cleaning effect of fingerprints improves with the increased hydrophilicity of the polyacrylate samples. This improvement can be attributed to the composition of fingerprints, which primarily consist of water, salt, and grease. The hydrophilicity of the polyacrylate samples plays a crucial role in the penetration of water and salt contamination. Additionally, as the polyacrylate samples possess some degree of lipophilicity, they can effectively wet oils and fats present in the fingerprints. Consequently, polyacrylate samples such as PMLA and PMLEA-1, which exhibit low hydrophilicity, are expected to perform poorly in the removal of fingerprints.

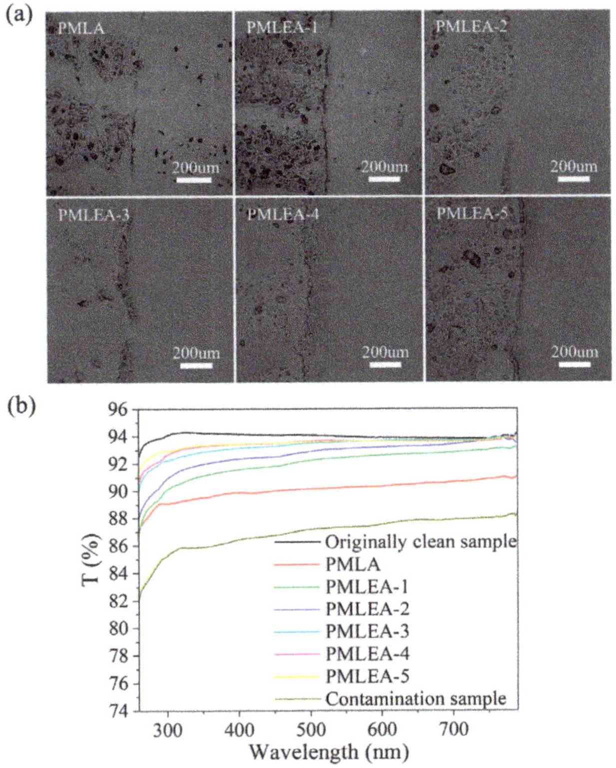

Figure 10. Comparison of the cleaning effect for removing fingerprints from the quartz glass. (**a**) the micrographs of the contrast areas and (**b**) the transmission spectra of the quartz glass before and after cleaning.

4. Conclusions

In this study, polyacrylates possessing both hydrophilic and lipophilic properties were synthesized via free radical polymerization. The structural elucidation and determination of molecular weight were conducted through FTIR, NMR, and GPC characterizations. The investigation into the polyacrylate samples' HLB, surface tension, and wettability allowed for an assessment of the impact of the LA to ER-10 ratio on the hydrophilicity, lipophilicity, and decontamination efficiency of the synthesized polyacrylates. Furthermore, the coating properties of the polyacrylate samples were found to be influenced by the proportions of LA and ER-10. Through comparative testing, it was revealed that the polyacrylate sample PMLEA-2 exhibited effectiveness in cleaning three common types of surface contaminants typically found on optical instruments. Additionally, the films derived from PMLEA-2 demonstrated balanced mechanical properties and ease of peeling, thereby effectively protecting the surface of the optics. Meanwhile, our results suggest that due to the diverse nature of different contaminants, it is recommended to use a combination of two different coatings to achieve the optimal cleaning effect. Most commercial peelable cleaning coatings use nitrocellulose as the film-forming material. In contrast, polyacrylate-based coatings offer enhanced stability and safety. Meanwhile, the Red First Contact Polymer product from Photonic Cleaning Technologies, which uses an inert polymer as the film-forming material, demonstrates superior cleaning performance and solvent formulation. Our coating exhibited overall decent performance but fell short in both of these aspects, highlighting the need for further refinement and improvement.

Supplementary Materials: The following supporting information can be downloaded at: https://www.mdpi.com/article/10.3390/ma17194813/s1, Figure S1: Preparation process of polyacrylates. Figure S2: (a) FTIR spectrum of ER-10, (b) ^{1}H-NMR spectrum of ER-10. Figure S3: ^{1}H-NMR spectra of the copolymer samples. Figure S4: GPC chromatograms of the copolymers. Table S1: Areas of the characteristic peaks.

Author Contributions: X.L., F.D. and L.S. conceived of the idea and supervised the research; D.Z. synthesized the samples and conducted the characterizations; D.Z., H.H., A.L., Y.Y., B.H. and A.A. conducted the characterizations, analyzed the data, and optimized the system; D.Z. and F.D. wrote the first draft of the manuscript, and all the authors contributed to the revision the manuscript. All authors have read and agreed to the published version of the manuscript.

Funding: This research was funded by the Fujian Provincial Industry-University Corporation Research Project (Grant no. 2022H6023).

Institutional Review Board Statement: Not applicable.

Informed Consent Statement: Not applicable.

Data Availability Statement: All the data generated or analyzed during this study are included in this published article and its supplementary information files.

Conflicts of Interest: The authors declare that there are no conflicts of interest.

References

1. Ling, X.L.; Zhao, Y.N.; Li, D.W.; Shao, J.D.; Fan, Z.X. Damage investigations of AR coating under atmospheric and vacuum conditions. *Opt. Laser Technol.* **2009**, *41*, 857–861. [CrossRef]
2. Li, X.G.; Shen, J. The stability of sol–gel silica coatings in vacuum with organic contaminants. *J. Sol-Gel Sci. Technol.* **2011**, *59*, 539–545. [CrossRef]
3. Cui, Y.; Zhao, Y.N.; Yu, H.; He, H.B.; Shao, J.D. Impact of organic contamination on laser-induced damage threshold of high reflectance coatings in vacuum. *Appl. Surf. Sci.* **2008**, *254*, 5990–5993. [CrossRef]
4. Yang, L.; Xiang, X.; Miao, X.X.; Li, Z.J.; Li, L.; Yuan, X.D.; Zhou, G.R.; Lv, H.B.; Zu, X.T. Influence of oil contamination on the optical performance and laser induced damage of fused silica. *Opt. Laser Technol.* **2015**, *75*, 76–82. [CrossRef]
5. Bude, J.; Carr, C.W.; Miller, P.E.; Parham, T.; Whitman, P.; Monticelli, M.; Raman, R.; Cross, D.; Welday, B.; Ravizza, F. Particle damage sources for fused silica optics and their mitigation on high energy laser systems. *Opt. Express* **2017**, *25*, 11414–11435. [CrossRef]
6. Ristau, D.; Jupé, M.; Starke, K. Laser damage thresholds of optical coatings. *Thin Solid Films* **2009**, *518*, 1607–1613. [CrossRef]
7. Lielienfeld, P. Optical Detection of Particle Contamination of Surface: A Review. *Aerosol Sci. Technol.* **1986**, *5*, 145–165. [CrossRef]
8. Ye, Y.Y.; Yuan, X.D.; Xiang, X.; Cheng, X.F.; Miao, X.X. Laser cleaning of particle and grease contaminations on the surface of optics. *Optik* **2012**, *123*, 1056–1060. [CrossRef]
9. Bennett, J.M. How to clean surfaces. In *Laser-Induced Damage in Optical Materials, Proceedings of the 2003 (XXXV Annual Symposium on Optical Materials for High Power Lasers: Boulder Damage Symposium), Boulder, CO, USA, 22–24 September 2003*; SPIE: Philadelphia, PA, USA, 2003; Volume 5273, pp. 195–206.
10. Bennett, J.M. Dos and don'ts in characterizing and cleaning optical surfaces. In *Optical Fabrication, Testing, and Metrology II, Proceedings of the Optical Systems Design 2005, Jena, Germany, 12–16 September 2005*; SPIE: Philadelphia, PA, USA, 2005; Volume 5965, pp. 392–401.
11. Houston, K.D. Comparative mirror cleaning study: A study on removing particulate contamination. In *Optical Systems Degradation, Contamination, and Stray Light: Effects, Measurements, and Control II, Proceedings of the SPIE Optics + Photonics, San Diego, CA, USA, 13–17 August 2006*; SPIE: Philadelphia, PA, USA, 2006; Volume 6291, pp. 61–68.
12. Kimura, W.D.; Kim, G.H.; Balick, B. Comparison of laser and CO_2 snow cleaning of astronomical mirror samples. In *Advanced Technology Optical Telescopes V, Proceedings of the 1994 Symposium on Astronomical Telescopes and Instrumentation for the 21st Century), Kailua-Kona, HI, USA, 13–18 March 1994*; SPIE: Philadelphia, PA, USA, 1994; Volume 2199, pp. 1164–1171.
13. Máša, V.; Horňák, D.; Petrilák, D. Industrial use of dry ice blasting in surface cleaning. *J. Clean. Prod.* **2021**, *329*, 129630. [CrossRef]
14. Lobmeyer, L.; Carey, L. Optical cleaning to remove particles for JWST mirror surfaces. In *Systems Contamination: Prediction, Control, and Performance 2018, Proceedings of the SPIE Optical Engineering + Applications, San Diego, CA, USA, 19–23 August 2018*; SPIE: Philadelphia, PA, USA, 2018; Volume 10748, pp. 140–149.
15. Rigolet, F. Cleaning and surface preparation: Which lasers for which applications. *Assem. Autom.* **1999**, *19*, 104–108. [CrossRef]
16. Mann, K.; Wolff-Rottke, B.; Mu, F. Cleaning of optical surfaces by excimer laser radiation. *Appl. Surf. Sci.* **1996**, *96–98*, 463–468. [CrossRef]
17. Lizon, J.L.; Deiries, S. Cleaning of extremely sensitive optical surfaces. In *Advances in Optical and Mechanical Technologies for Telescopes and Instrumentation, Proceedings of the SPIE Astronomical Telescopes + Instrumentation, Montréal, QC, Canada, 22–27 June 2014*; SPIE: Philadelphia, PA, USA, 2014; Volume 9151, pp. 1793–1800.

18. Gossen, K.; Ehrmann, A. Influence of FTO glass cleaning on DSSC performance. *Optik* **2019**, *183*, 253–256. [CrossRef]
19. Barthel, A.J.; Luo, J.W.; Hwang, K.S.; Lee, J.-Y.; Kim, S.H. Boundary lubrication effect of organic residue left on surface after evaporation of organic cleaning solvent. *Wear* **2016**, *350–351*, 21–26. [CrossRef]
20. Dey, T.; Naughton, D. Cleaning and anti-reflective (AR) hydrophobic coating of glass surface: A review from materials science perspective. *J. Sol-Gel Sci. Technol.* **2016**, *77*, 1–27. [CrossRef]
21. Wagle, P.G.; Tamboli, S.S.; More, A.P. Peelable coatings: A review. *Prog. Org. Coat.* **2021**, *150*, 106005. [CrossRef]
22. Joseph, R. The Lowdown on Peelable Coatings Specs and Usage. *Met. Finish.* **2004**, *102*, 45–46. [CrossRef]
23. Gardon, J.L. Peel adhesion. II. A theoretical analysis. *J. Appl. Polym. Sci.* **1963**, *7*, 643–665. [CrossRef]
24. Liprandi, D.; Bosia, F.; Pugno, N.M. A theoretical-numerical model for the peeling of elastic membranes. *J. Mech. Phys. Solids* **2020**, *136*, 103733. [CrossRef]
25. Cipriano, J. Cleaning Modern Antireflection-Coated Optics with Collodion. Cloudy Nights Telescope Reviews (Blog). 25 November 2005. Available online: https://www.cloudynights.com/articles/cat/articles/how-to/collodion-optics-cleaning-r1265 (accessed on 21 March 2024).
26. Baker, M.I.; Walsh, S.P.; Schwartz, Z.; Boyan, B.D. A review of polyvinyl alcohol and its uses in cartilage and orthopedic applications. *J. Biomed. Mater. Res. Part B Appl. Biomater.* **2012**, *100*, 1451–1457. [CrossRef]
27. Rao, S.V.S.; Lal, K.B. Surface decontamination studies using polyvinyl acetate based strippable polymer. *J. Radioanal. Nucl. Chem.* **2004**, *260*, 35–42. [CrossRef]
28. Lewandowski, K.; Krepski, L.R.; Mickus, D.E. Dry-peelable temporary protective coatings from waterborne self-crosslinkable sulfourethane–silanol dispersions. *J. Appl. Polym. Sci.* **2004**, *91*, 1443–1449. [CrossRef]
29. Long, N.H.; Park, H.W.; Chae, G.S.; Lee, J.H.; Bae, S.W.; Shin, S. Preparation of peelable coating films with a metal organic framework (UiO-66) and self-crosslinkable polyurethane for the decomposition of methyl paraoxon. *Polymers* **2019**, *11*, 1298. [CrossRef]
30. He, Z.Y.; Li, Y.T.; Xiao, Z.Q.; Jiang, H.; Zhou, Y.L.; Luo, D.L. Synthesis and Preparation of (Acrylic Copolymer) Ternary System Peelable Sealing Decontamination Material. *Polymers* **2020**, *12*, 1556. [CrossRef]
31. Wang, X.R.; Ma, G.Y.; Ma, H.F.; Wu, Z. Stripping and mechanical properties of water-based polyacrylate stripper and its application in art masking fluid. *Iran. Polym. J.* **2021**, *30*, 25–34. [CrossRef]
32. Xu, X.R.; Li, J.; Pan, X.H.; Lin, X.Y.; Li, Z.G. Preparation of high tensile strength modified cellulose-based strippable coating and its application on radioactive decontamination at low-temperature. *Mater. Today Commun.* **2023**, *34*, 105088. [CrossRef]
33. Tang, J.L.; Li, S.Y.; Wang, Y.Y.; Zhang, H.L.; Lin, B.; Sun, M.X. Ethyl cellulose based peelable coatings with visual sensing of hydrogen sulfide. *Prog. Org. Coat.* **2022**, *163*, 106617. [CrossRef]
34. Kohli, R. Applications of Strippable Coatings for Removal of Surface Contaminants. In *Developments in Surface Contamination and Cleaning: Applications of Cleaning Techniques*; Elsevier: Amsterdam, The Netherlands, 2019; pp. 49–96.
35. Von Rybinski, W. Physical Aspects of Cleaning Processes. In *Handbook for Cleaning/Decontamination of Surfaces*; Elsevier: Amsterdam, The Netherlands, 2007; pp. 1–55.
36. Song, Y.; Zhou, G.Y.; Tu, M.T.; Zhang, J.C.; Wang, P. Effectiveness of oolong tea and simethicone solution for lens cleansing during colonoscopy: A double-blinded randomized study. *Medicine* **2021**, *100*, e26763. [CrossRef]
37. Michon, B.; López-Sánchez, U.; Degrouard, J.; Nury, H.; Leforestier, A.; Rio, E.; Salonen, A.; Zoonens, M. Role of surfactants in electron cryo-microscopy film preparation. *Biophys. J.* **2023**, *122*, 1846–1857. [CrossRef] [PubMed]
38. Anton, W.L.; Spinelli, H.J.; Patil, A.A. Acrylic Polymer Compounds. U.S. Patent 57,984,26A, 25 August 1998.
39. Yamashita, K.; Matsuki, M.; Asai, H.; Matsuyama, N.; Tojo, H.; Kurota, H.; Akasaka, K.; Obara, H. Aqueous Dispersion of a Peelable Coating Composition. U.S. Patent 634,423,6B2, 5 February 2002.
40. Pasquali, R.C.; Sacco, N.; Bregni, C. The studies on hydrophilic-lipophilic balance (HLB): Sixty years after William C. Griffin's pioneer work (1949–2009). *Lat. Am. J. Pharm.* **2009**, *28*, 313–317.
41. Griffin, W.C. Calculation of HLB values of non-ionic surfactants. *J. Cosmet. Sci.* **1954**, *5*, 24–31.
42. Greenwald, H.L.; Brown, G.L.; Fineman, M.N. Determination of hydrophile-lipophile character of surface active agents and oils by water titration. *Anal. Chem.* **1956**, *28*, 1693–1697. [CrossRef]
43. Ananthapadmanabhan, K.P. Surfactant Solutions: Adsorption and Aggregation Properties. In *Interactions of Surfactants with Polymers and Proteins*; CRC Press: Boca Raton, FL, USA, 1993; pp. 5–58.

materials

MDPI

Article

Effect of N_2/Ar Ratio on Wear Behavior of Multi-Element Nitride Coatings on AISI H13 Tool Steel

Cheng-Hsun Hsu *, Hong-Wei Chen, Chun-Yin Lin and Syue-Hong Hu

Department of Mechanical and Materials Engineering, Tatung University, Taipei City 10451, Taiwan; tonychen318@gmail.com (H.-W.C.); rupert475@gmail.com (C.-Y.L.); alex0979306954@gmail.com (S.-H.H.)
* Correspondence: chhsu@gm.ttu.edu.tw

Abstract: In this study, multi-element nitride coatings composed of (Ti, Cr, Cu, Al, Si)N were synthesized on H13 tool steel using cathodic arc deposition (CAD) technology. The N_2/Ar flow ratio varied from 0 to 2 as the experimental parameter, and two targets, Ti-Cr-Cu and Al-Si alloys, were utilized simultaneously. The impact of the gas flow ratio on the coatings' abrasion properties was investigated, focusing on aspects, such as chemical composition, adhesion, hardness, and wear behavior. The experimental findings indicate that the coated specimens with a nitrogen reaction exhibit superior hardness and abrasion resistance compared to those without nitrogen use. While the surface roughness of the specimens tends to increase slightly with a higher N2/Ar ratio, the coating demonstrates improved hardness, adhesion, and abrasion resistance performance. In summary, the wear-resistant characteristics of H13 tool steel can be significantly enhanced when applying a CAD-(Ti, Cr, Cu, Al, Si)N film with a flow ratio of N_2/Ar = 2.

Keywords: H13 tool steel; cathodic arc deposition; N_2/Ar flow ratio; (Ti, Cr, Cu, Al, Si)N coating; adhesion; wear resistance

1. Introduction

AISI H13 steel is widely recognized as a hot work tool steel, renowned for its composition comprising chromium (Cr), molybdenum (Mo), vanadium (V), and other alloying elements. This composition imbues H13 steel with exceptional toughness, abrasion resistance, and high-temperature fatigue resistance, rendering it suitable for prolonged usage in elevated temperature conditions. Additionally, its ability to withstand temperature fluctuations without significant deformation further underscores its utility. H13 steel finds extensive application in various industries, particularly in the fabrication of hot-forming molds, stamping molds, die-casting molds, and hot work tools [1–3]. However, due to cost considerations in the industry, the same die-casting mold or extrusion mold is usually used repeatedly. After a long period, the soldering phenomenon appears on the surface. This is the main reason for reducing the steel's service life, so various surface treatments are often used to improve the tool steel's service life. For example, Yaqoob et al. [4] used Al_2O_3/TiCN-CVD and TiAlN/AlCrN-PVD multilayer films to coat tool steel to improve the service life of cutting high-strength steel. Varis et al. [5] used heat treatment to enhance the wear resistance of HVAF and HVOF spray tool coatings. Poirier et al. [6] also used heat treatment to improve the ability of tool steel powder cold spraying. Temmler et al. [7] discussed the effect of laser micro polishing on the surface roughness of AISI D2 tool steel, and Ormanova et al. [8] used electron beam surface treatment to improve the hardness of tool steel to 718 kg/mm^2.

The physical vapor deposition (PVD) method has been widely used in the field of industrial hard coatings in recent decades. The process of the PVD method is mainly carried out in a low-temperature vacuum reactor, using gas discharge to ionize the evaporated metal and the incoming gases (such as nitrogen, acetylene, etc.), as well as reacting to generate nitrides or carbides on the surface of the workpiece. Nitrides and carbides have very high hardness and generally exhibit strong adhesion, making them suitable for applications requiring wear resistance. In practical applications, there are many types of PVD-coated hard films, such as TiN [9,10], CrN [10–12], ZrN [13,14], CrAlN [15], Ti-CuN [16], TiAlN [17,18], and other binary/ternary nitrides [19–22] are mainly used in surface coatings of cutting tools and molds. They often have the advantages of high surface hardness and low friction coefficient.

Additionally, there is the possibility of applying a quaternary coating, such as a CrAlSiN composite ceramic film, which can be manufactured using PVD technology [23–27]. The related reports indicate that the film exhibits exceptional properties, including high hardness, excellent friction, and oxidation resistance. The wear resistance of CrAlSiN is superior to that of CrAlN, and the presence of silicon in the CrAlSiN film helps to lower its friction coefficient at room temperature. Chang et al. [26] used the cathodic arc vapor deposition method to coat TiAlN, CrAlSiN, and TiAlSiN films on tungsten carbide milling tools. The resulting TiAlN, CrAlSiN, and TiAlSiN had ultra-high hardness values of 31 ± 1 GPa, 36 ± 2 GPa, and 35 ± 2 GPa, respectively; the use life of CrAlSiN coated tools exceeded that of TiAlSiN- and TiAlN-coated tools [27]. This is because the CrAlSiN coating has a high wear resistance. Therefore, it can be seen that the nitride of multi-alloying element types is similar to that of high entropy alloys (HEAs) [28–30], which may have excellent properties as well as potential applications.

Therefore, this study employed the cathodic arc deposition (CAD) technique to coat H13 tool steel substrates with multi-elemental nitrides (Ti, Cr, Cu, Al, Si)N. The films were synthesized using co-reacting Ti-Cr-Cu and Al-Si dual targets, with variations in the N_2/Ar gas flow ratio. Subsequently, the impact of the N_2/Ar gas flow ratio on the wear resistance of H13 tool steel coated with this film was explored, aiming to enhance the longevity of the steel in service. Such research could provide a novel contribution to surface modification techniques for H13 tool steel.

2. Materials and Methods

2.1. Material Preparation and Pre-Coating Treatment

AISI H13 hot work tool steel substrates with a diameter of 25 mm and 6 mm thickness were used in this study, and their chemical composition was analyzed using glow discharge optical spectroscopy (GDOS), as listed in Table 1.

Table 1. Chemical compositions of the H13 steel used in the study (wt.%).

Cr	Mo	V	Cu	C	Si	Mn	P	S	Fe
5.36	1.31	0.92	0.09	0.38	0.94	0.47	0.02	0.03	Bal.

Before undergoing the coating procedure, the specimens were heated to form a high-hardness matrix suitable for practical applications. Specifically, a salt bath furnace was employed for this heat treatment process. The specimens were initially preheated at 550 °C for 15 min, and then further heated to 1050 °C for 30 min. Subsequently, they were quenched in water and tempered at 550 °C for 60 min, repeated twice. The outcome of the heat treatment revealed a significant increase in hardness, from the initial 13 HRC to 52 HRC.

It is a well-known fact that the roughness and cleanliness of the specimen's surface play pivotal roles in determining the coating quality and adhesion. As such, the specimens underwent a series of preparatory steps including grinding, polishing, and cleaning before the coating process. The details were as follows: sandpaper was utilized to grind the surface of the specimens, followed by polishing using 0.3 μm aluminum oxide with deionized water until a mirror-like finish was achieved. Subsequently, the specimens were immersed in an acetone solution and subjected to ultrasonic shock for 10 min to ensure thorough cleaning. Upon completion of the cleaning process, the specimens were dried in an oven set at 70 °C for 15 min. This meticulous preparation ensured that the surfaces of all specimens were devoid of any contaminants, thus readying them for the coating process within the CAD chamber.

2.2. CAD Treatment

The coating process is outlined as follows: the specimens were positioned inside the CAD chamber, which was then evacuated to a high vacuum level of 5×10^{-5} torr. Subsequently, argon was introduced to regulate the pressure at 2×10^{-2} torr. A bias voltage of −700 V was applied, and argon ion bombardment was conducted for 10 min to further cleanse the specimens' surfaces. Following the argon ion bombardment, the deposition process commenced. Initially, the Al-Si target was subjected to arcing for 5 min, resulting in the deposition of an AlSi intermediate layer on the H13 specimens. Subsequently, both the Ti-Cr-Cu and Al-Si targets were simultaneously activated, initiating the synthesis of films. To investigate the effect of N_2 reactive gas, the N_2/Ar flow ratios were manipulated as experimental parameters, specifically set at 0, 1, and 2, the three conditions in the study being numbered S0, S1, and S2 respectively. The coating process was concluded within 40 min, with the key process parameters listed in Table 2.

Table 2. CAD processing parameters for the coatings in this study.

Parameter	Value
Two targets	(Ti34%-Cr34%-Cu32%) and (Al50%-Si50%)
N_2/Ar flow ratio	(S0) N_2(0 sccm)/Ar(180 sccm) = 0 (S1) N_2(90 sccm)/Ar(90 sccm) = 1 (S2) N_2(120 sccm)/Ar(60 sccm) = 2
Cathode current (A)	60
Working pressure (Torr)	2×10^{-2}
Ar ion bombardment (V)	−700
Substrate bias (V)	−100
Substrate temperature (°C)	250
Rotation rate (rpm)	4
Distance between target and substrate (mm)	150
Total deposition time (min)	45

2.3. Coating Characteristics and Morphological Observation

In the study, various characteristics of the films were analyzed as follows. (1) The measurement of coating roughness was carried out with a surface roughness tester (Mitutoyo Surf test SV-400, Kawasaki, Japan) to obtain the average roughness (Ra value). (2) The chemical composition of the films was analyzed using a field emission electron probe micro-analyzer (EPMA, JEOL JXA-8530F Plus, Tokyo, Japan). (3) An X-ray diffractometer (Bruker D8 X-ray Powder Diffractometer, Karlsruhe, Germany) was used to identify the coating structure with Cu-target Kα radiation (λ = 1.54060 Å) at 40 kV and 30 mA, at a glancing incident angle of 2°; the scanning angle (2θ) ranged from 20° to 80° at a constant speed of 2°/min. (4) The adhesion strength quality (ASQ) of the coatings was evaluated by using Rockwell-C indentation testing with a load of 1471 N. The damage to the coatings was compared on a defined ASQ basis, where HF1–HF4 represented acceptable adhesion and HF5–HF6 represented insufficient adhesion (HF is the German short form for adhesion strength) [31]. (5) Nano-indentation measurements were performed using a nanoindenter (Hysitron TI 980 TriboIndenter, Hysitron, MN, USA) with a diamond Berkovich tip. According to the Oliver and McHargue method [32], both the values of hardness (H) and elastic modulus (E) for the coatings were measured by analyzing load–displacement curves under a load of 20 mN. The averages of 20 indentations for each film were reported with the standard deviation of both the H and E values. (6) A transmission electron microscope (TEM, Philips TECHNAIG-2F20, Cambridge, MA, USA) was used to observe multilayer morphology in the film, and a field emission scanning electron microscope (FE-SEM, HITACHI SU-8000, Tokyo, Japan) was used to measure the thickness of the coatings. (7) Also, a micro-hardness tester was used to measure the surface hardness of the coated specimens. A load of 300 g was applied in the micro-hardness tests. Each specimen was tested three times to obtain an average HV value.

2.4. Wear Test

Based on previous wear test experiences [10,33], the wear tests were performed using a ball-on-disk tribometer (CSM Instruments, Peseux, Switzerland). The station view of the friction node used in the wear tests is shown in Figure 1. The testing conditions were as follows: (1) no lubricant, (2) circular track with a 5 mm radius against a 6 mm diameter WC-6%Co ball, (3) wear speed of 0.17 m/s, (4) a load of 2 N, (5) at room temperature, and (6) 35% relative humidity. The relationship between the friction coefficient and the total travel distance of 1200 m was continuously recorded during the testing. The wear rate was derived from the weight loss of the specimens, which was evaluated using a microbalance with an accuracy of 1×10^{-4} g. Three tests were performed for each coating condition, and the average wear rate was calculated. Furthermore, the worn surface of the specimens after the wear testing was observed using the FE-SEM.

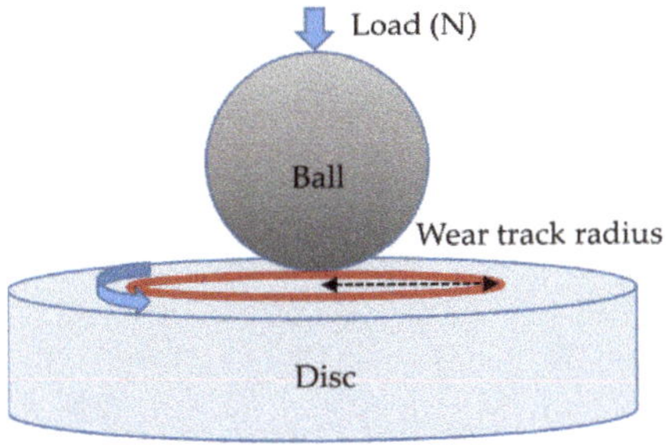

Figure 1. Schematic of the friction node used in the wear tests. The red circle is the wear track, and the dotted arrow is the wear track radius.

3. Results and Discussion

3.1. Coating Composition and Structure

The resultant composition of the three coatings analyzed by EPMA is listed in Table 3, and the structure of the films was also examined using XRD, as shown in Figure 2. It can be found that the S0 film was synthesized without nitrogen addition, the content of Cu and Si was smaller as compared to Ti, Cr, and Al elements. The constitute phases in the coating are mainly composed of intermetallic compounds such as Cr_2Ti, Cu_4Ti, and AlSi. When nitrogen gas was added to the S1 specimen ($N_2/Ar = 1$), the contents of Cu and Si in the coating were still low, and the contents of Ti and Cr were also reduced, while the content of Al was the highest. It could be found that the characteristic peaks of the XRD diffraction pattern appeared at $2\theta = 44.7°$ for AlN(200), $2\theta = 40.1°$ for TiN(110), and $2\theta = 38.0°$ for CrN(111). Therefore, the coating structure of S1 contained a small amount of AlN, TiN, and CrN, attributed to the presence of nitrogen, alongside the formation of intermetallic compound phases, as observed in the S0 specimen. When the N_2/Ar ratio reached 2 (S2), the nitrogen content in the coating increased significantly. As a result, nitrides such as AlN, CrN, and TiN became predominant in the coating structure, whereas the presence of intermetallic compounds was notably reduced. In addition, it can be seen that the AlN phase shows the strongest peak in the XRD pattern because of the highest content of Al and N in the S2 film (see Table 3). Due to the presence of a small amount of silicon, Si doping in AlN results in the formation of a minor quantity of AlSiN. The difference in the three coating structures may affect the abrasion properties of the coated specimens as discussed in the subsequent abrasion results.

Table 3. Chemical composition of the films analyzed by EPMA (at.%).

Specimen	Ti	Cr	Cu	Al	Si	N
S0	26.25 ±1.31	32.46 ±1.66	8.54 ±0.42	27.99 ±1.39	4.76 ±0.23	0
S1	18.51 ±0.92	21.62 ±1.08	5.85 ±0.29	28.53 ±1.42	3.38 ±0.17	22.11 ±1.10
S2	11.26 ±0.56	10.41 ±0.52	3.02 ±0.15	28.77 ±0.14	3.42 ±0.17	43.12 ±2.15

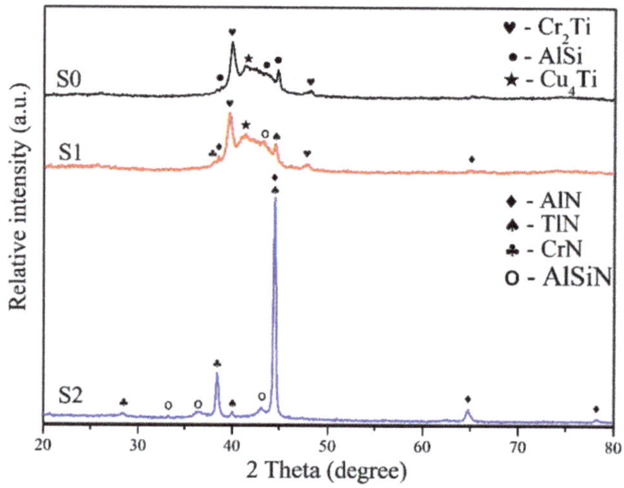

Figure 2. XRD patterns of the coatings in the study.

3.2. Coating Morphology and Roughness

Figure 3 depicts the surface morphology of three coated specimens, revealing many micro-droplets on their surfaces. Figure 3a shows the S0 film without nitrogen, it seems that the size and quantity of the micro-droplets on the surface are smaller compared to the S1 and S2 films. Further comparing the difference between S1 and S2 (Figure 3b,c), it was also found that the droplet sizes on the S1 specimen seemed to be slightly smaller than those on the S2 specimen. To determine the coating thickness, the cross-sections of the coated specimens were examined using FE-SEM. Figure 4a–c show the cross-sectional morphologies of the three coated specimens, each with an average film thickness of approximately S0 = 8.06 μm, S1 = 8.38 μm, and S2 = 9.67 μm, respectively. This result also shows that the film thickness increases as the N_2/Ar ratio increases. Figure 5 shows a high-magnification TEM cross-section of the S2 film. It can been seen there is a clear black and white nano-scale multilayered structure formed in the film. The reason is that the specimen carrier has a rotation rate of 4 rpm during the coating process. The white layer is approximately 17 nm thick and is (Al, Si)N, while the black layer about 10 nm thick and is (Ti, Cr, Cu)N. A greater thickness of the (Al, Si)N layer suggests that AlN forms more rapidly than the other nitrides at the same rotation speed, resulting in a thicker layer in the multilayer film. Moreover, the (Al, Si)N layer is composed of AlN mixed with a small amount of AlSiN phase.

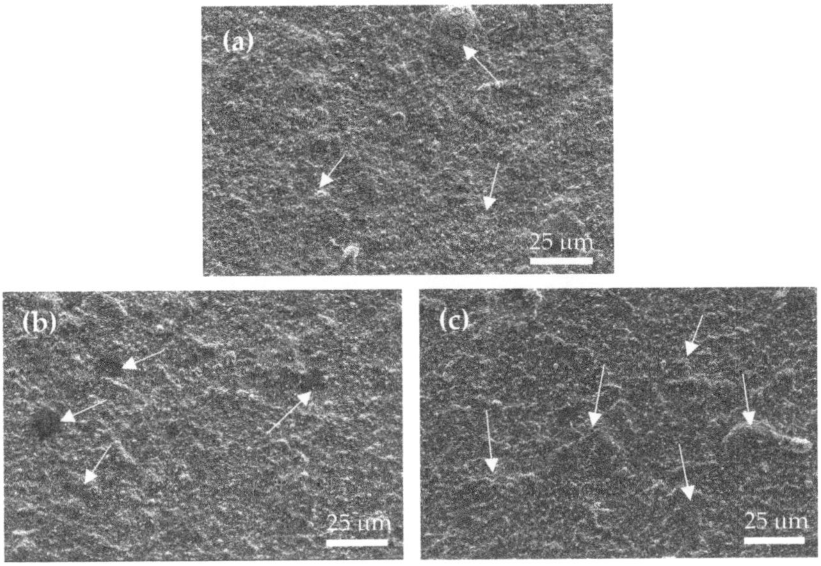

Figure 3. Surface morphology of the coated specimens. (**a**) S0, (**b**) S1, and (**c**) S2. The arrow points to the distribution of microdroplets.

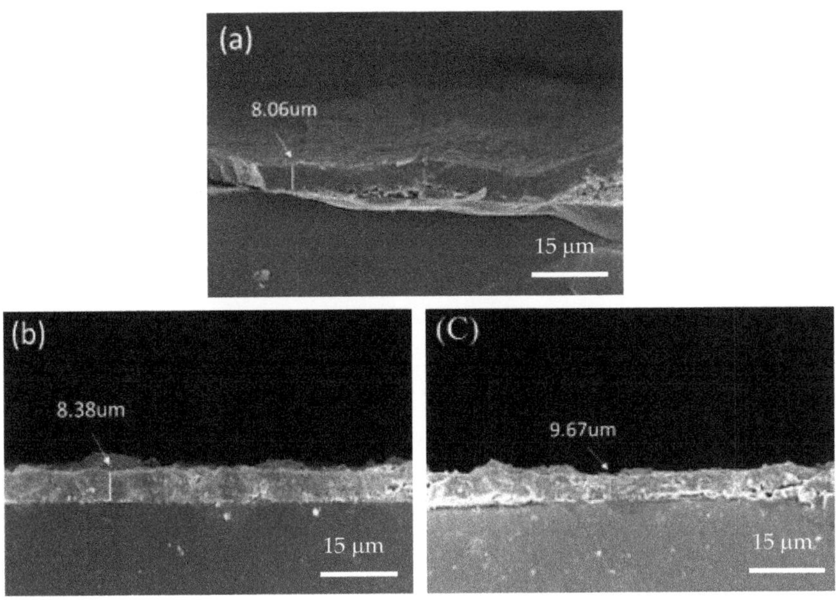

Figure 4. Cross-sectional view of the coated specimens: (**a**) S0, (**b**) S1, and (**c**) S2.

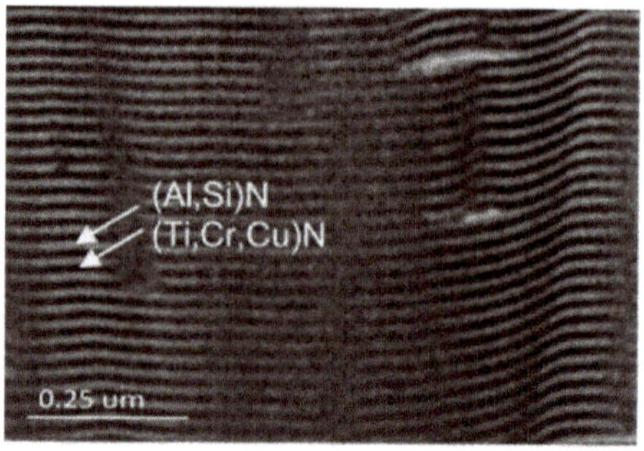

Figure 5. TEM cross-sectional appearance of the S2 multilayered film.

The surface roughness of the film layer is mainly affected by the sputtering of droplets on the surface of the coated specimen during the cathodic arc deposition process [34], i.e., the generation of droplets is due to the sputtering of larger neutral particles and atomic groups when the arcing position of the target is in the high-temperature melting state. In this study, the measurement of coating roughness was performed to obtain the average roughness (Ra value). The testing principle of surface roughness is to use a probe to scan the ups and downs of the surface of the object to make the probe vibrate vertically, and then collect the vertical vibration signal of the probe through a computer to achieve the measurement result. Because the Ra value is the average surface roughness, it is often used to represent the surface roughness of materials and is the most representative. Figure 6 illustrates the relationship between the surface roughness (Ra value) of each specimen before and after coating. The comparison reveals that the coated specimens exhibit higher Ra values, which can be attributed to microdroplets formed during the CAD process. Additionally, it is observed that the surface roughness increases with a higher N_2/Ar ratio, resulting in S2 having a greater Ra value compared to S1.

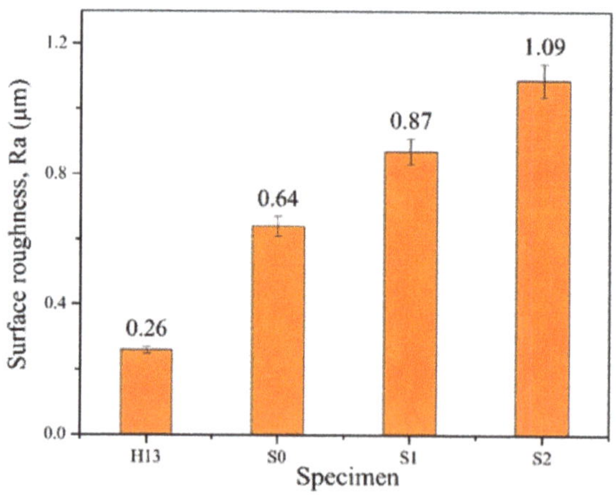

Figure 6. Surface roughness of the uncoated and coated specimens.

3.3. Coating Adhesion and Hardness

In this study, the German specification VDI 3198 was used to determine the adhesion of the coating [31], i.e., an HRC indentation test of the film layer using Rockwell hardness machine, followed by observation of the rupture periphery of the film layer using SEM, and classification of the indentation in terms of adhesion according to the classification of rupture degree in the specification. Figure 7 shows the photographs of the indentation rupture on the surfaces of three coated specimens with arrows pointing to film peeling around the indentation. From Figure 7a, it can be found that the flaking of the S0 film without nitrogen is severe, and there is obvious film peeling around the indentation, so the S0 specimen can be assumed as HF4. Figure 7b is S1, and the graph shows that there is a little flaking around the indentation, but the extent of the flaking is lower than that of the S0 specimen, and therefore S1 is judged to be HF3. Figure 7c is S2, and this film almost has no obvious cracking and flaking. It has the best performance in terms of adhesion among the coated specimens; therefore, S2 can be classified as HF2. Therefore, the adhesion is S2 > S1 > S0 from the best to the worst among the three coated specimens.

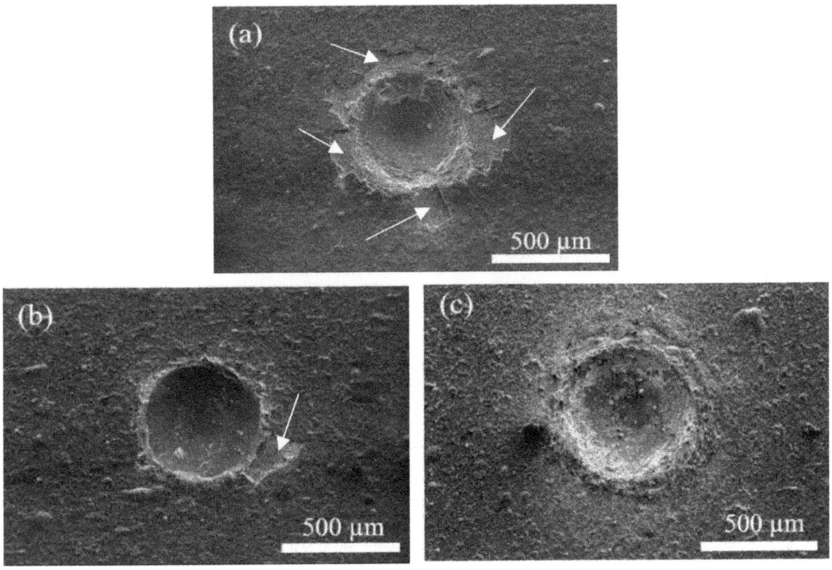

Figure 7. Adhesion comparison of the coated specimens. (**a**) S0, (**b**) S1, and (**c**) S2. The arrows point to the location where the film peels off.

To understand the effect of N_2/Ar flow ratio on film hardness and avoid the occurrence of substrate effect, this study used nano-indentation to accurately measure the film hardness (H) and its elastic modulus (E). Figure 8a–c shows the hardness (H), modulus of elasticity (E), and H/E values of the three types of coated specimens. The experimental results demonstrated that the hardness values were highest for S2 at 17.5 GPa, followed by S1 at 14.1 GPa, and lowest for S0 at 8.8 GPa. There is a notable difference in hardness between S0 and the other two specimens, S1 and S2. This disparity is attributed to the absence of nitride in the S0 film, resulting in its lower hardness. In contrast, both S1 and S2 films contain nitride, contributing to their higher hardness values. Particularly, the S2 film exhibited the highest hardness due to its greater nitride content compared to the S1 film. Regarding the H/E values, the elastic modulus (E) for the three coated specimens are as follows: S0 = 138.3 GPa, S1 = 198.7 GPa, and S2 = 115.9 GPa. Using these E values, the H/E ratios can be calculated. Comparing the H/E ratios of the three coatings, the ranking is S2 > S1 > S0.

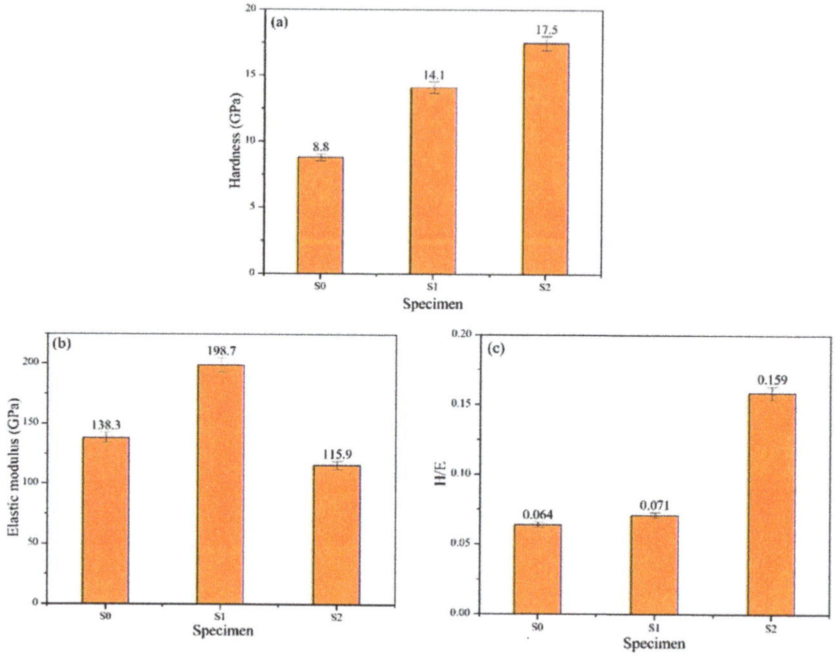

Figure 8. Comparison of (**a**) H, (**b**) E, and (**c**) H/E values for the coated specimens.

Additionally, Figure 9 illustrates a comparison of HV hardness values for the uncoated and coated specimens. The data indicate that the surface hardness of the uncoated H13 substrate, which measures 526 HV, can be substantially improved through coating. Among the coated specimens, S2 exhibits the highest hardness value at 1857 HV, followed by S1 at 1522 HV, and S0 with the lowest hardness value of 1354 HV. This hardness relationship ranking aligns with the previously discussed trend in nano-hardness variation (Figure 8a). Therefore, an increase in the amount of nitrides formed during coating markedly boosts the surface hardness of the H13 substrate.

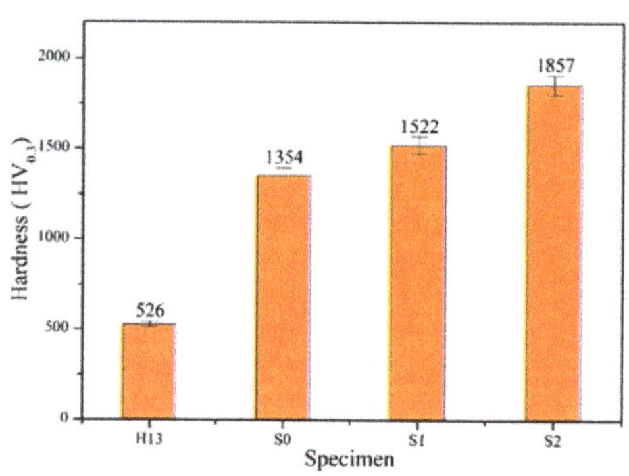

Figure 9. Comparison of Vickers hardness values among H13 and the coated specimens.

3.4. Analysis of Wear Behavior

Some literature [35–37] pointed out that a higher H/E ratio in coatings is linked to reduced accumulated strain within the coating layer, thereby improving wear resistance. In other words, a higher H/E value is associated with better abrasion resistance. Therefore, the wear behavior of a coating material is affected by the H/E value. Figure 10 shows the correlation between the friction coefficient and the wear distance for H13 steel with and without coating after the 1200 m wear test. The curve in the figure shows that the friction coefficient of the uncoated H13 substrate seems to have a slightly rising trend, so its friction coefficient is slightly high. As a result, the uncoated substrate seems to have a higher friction coefficient among all the specimens. Further comparing the difference among the three coated specimens, we also find that after the wear distance of 800 m, the curves of friction coefficients for both the S0 and Si films are slightly close to each other to show a similar friction coefficient, while the S2 film has the lowest friction coefficient. This is because S2 has the most nitride content to show the highest hardness as well as the lowest friction coefficient. According to the curves in Figure 10, a comparison of the wear rate for each specimen can be further obtained, as shown in Figure 11. The comparison of data shows that the uncoated specimen has the highest wear rate, while the coating performance can reduce its wear rate. S2 has the lowest wear rate, so it has the best performance for improving the wear resistance. This result also matches those mentioned above, i.e., the abrasion behavior of the coating has a direct correlation with the H/E value. S2 has the highest H/E value to show optimal wear resistance. Figure 12 displays the SEM wear trajectory of the uncoated and coated specimens. From Figure 12a, it can be observed that the uncoated specimen after the wear test shows an apparent scratch and the wear width is the largest (486 μm), followed by the S0 and S1 specimens as seen in Figure 12b,c, which are 464 μm and 442 μm, respectively, while S2 has the narrowest wear width of 392 μm (Figure 12d). Therefore, it conforms to the comparison results in Figure 11, which show that S2 has the best wear resistance.

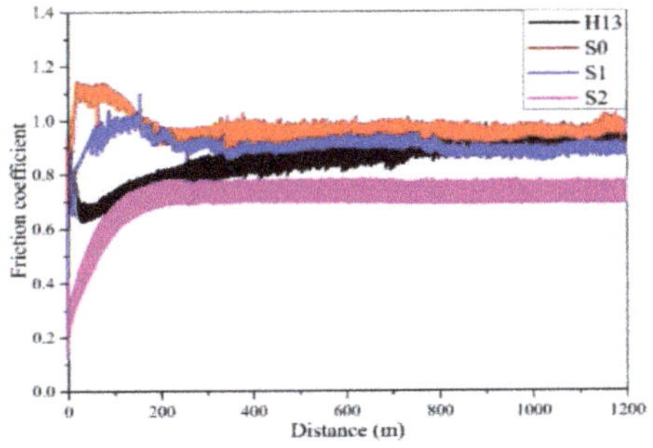

Figure 10. Friction coefficient curves of H13 and the coated specimens.

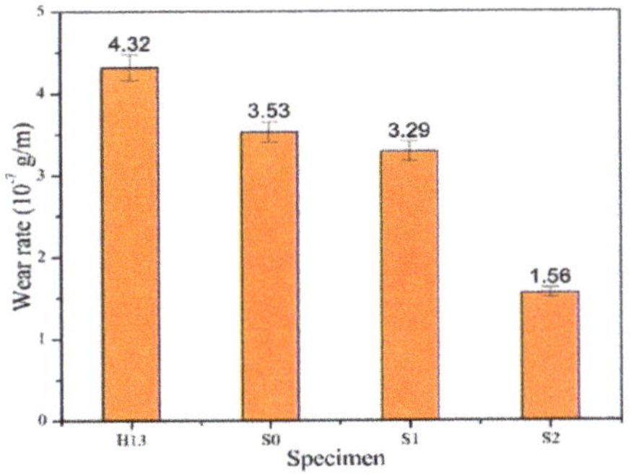

Figure 11. Comparison of wear rates among H13 and the coated specimens.

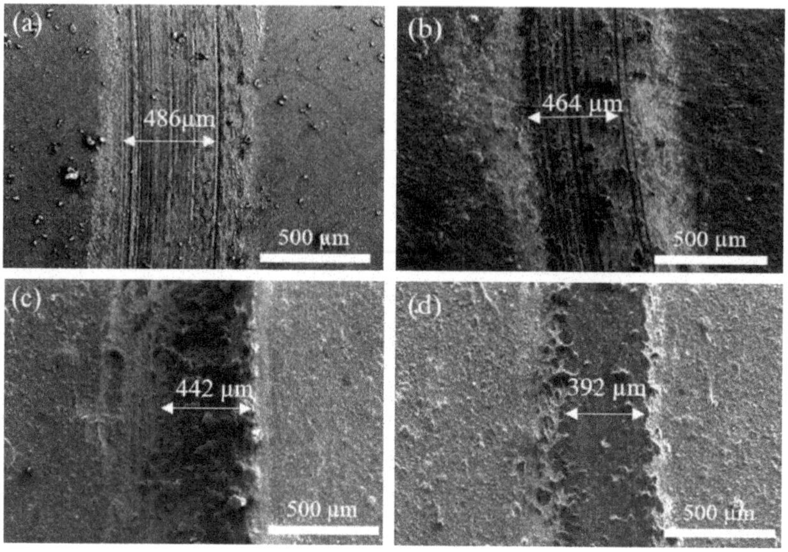

Figure 12. SEM surface morphologies of the uncoated and coated specimens after wear test: (**a**) H13, (**b**) S0, (**c**) S1, and (**d**) S2.

4. Conclusions

1. This study employed cathodic arc deposition to successfully coat (Ti, Cr, Cu)N and (Al, Si)N nano-scale interleaved multilayers on H13 steel, with good adhesion of the coating to the substrate.
2. In the absence of N_2, the film structure consists primarily of Cr_2Ti, Cu_4Ti, and AlSi intermetallic compounds. When the N_2/Ar ratio is set to 2, the film structure predominantly features AlN, TiN, and CrN nitrides, with a minor presence of AlSiN.
3. As the N_2/Ar ratio increases, the hardness of the film rises, while the surface roughness of the coating also increases.
4. In terms of wear behavior, the wear resistance of H13 steel can be enhanced by applying a CAD-(Ti, Cr, Cu, Al, Si)N multilayer coating. The best wear resistance is achieved with a N_2/Ar ratio of 2.

Author Contributions: Conceptualization, C.-H.H.; Methodology, C.-H.H.; Validation, C.-Y.L.; Formal analysis, H.-W.C. and C.-Y.L.; Investigation, C.-H.H., H.-W.C., C.-Y.L. and S.-H.H.; Data curation, H.-W.C. and S.-H.H.; Writing—original draft, C.-H.H.; Writing—review & editing, C.-H.H. All authors have read and agreed to the published version of the manuscript.

Funding: The authors would like to thank National Science and Technology Council (Taiwan): NSTC113-2221-E-036-002 and Tatung University (Taipei City, Taiwan): B113-M05-015 for their funding support.

Institutional Review Board Statement: Not applicable.

Informed Consent Statement: Not applicable.

Data Availability Statement: The original contributions presented in the study are included in the article, further inquiries can be directed to the corresponding author.

Conflicts of Interest: The authors declare no conflict of interest.

References

1. Meng, Y.; Sugiyama, S.; Yanagimoto, J. Microstructural evolution during RAP process and deformation behavior of semi-solid SKD61 tool steel. *J. Mater. Process. Technol.* **2012**, *212*, 1731–1741. [CrossRef]
2. Shin, S.S.; Kim, Y.I.; Kim, K.H.; Lee, B. Microstructure and mechanical properties of SKD61–Cu cooling channel fabricated via explosive welding. *Mater. Today Comm.* **2021**, *28*, 102645. [CrossRef]
3. Yan, B.H.; Tsai, F.C.; Sun, L.W.; Hsu, R.T. Abrasive jet polishing on SKD61 mold steel using SiC coated with Wax. *J. Mater. Process. Technol.* **2008**, *208*, 318–329. [CrossRef]
4. Yaqoob, S.; Ghani, J.A.; Jouini, N.; Juri, A.Z. Performance evaluation of PVD and CVD multilayer-coated tools in machining high-strength Steel. *Coatings* **2024**, *14*, 865. [CrossRef]
5. Varis, T.; Lagerbom, J.; Suhonen, T.; Raami, L.; Terho, S.; Laurila, J.; Peura, P.; Vuoristo, P. Effect of heat treatments on the wear resistance of HVAF and HVOF sprayed tool steel coatings. *Surf. Coat. Technol.* **2023**, *462*, 129508. [CrossRef]
6. Poirier, D.; Thomas, Y.; Guerreiro, B.; Martin, M.; Aghasibeig, M.; Irissou, E. Improvement of tool steel powder cold spray ability via softening and agglomeration heat treatments. *J. Therm. Spray Technol.* **2022**, *31*, 145–158. [CrossRef] [PubMed]
7. Temmler, A.; Cortina, M.; Ross, I.; Küpper, M.E.; Rittinghaus, S.K. Laser micro polishing of tool steel 1.2379 (AISI D2): Influence of intensity distribution, laser beam size, and influence on surface roughness and area rate. *Metals* **2021**, *11*, 1445. [CrossRef]
8. Ormanova, M.; Petrov, P.; Kovacheva, D. Electron beam surface treatment of tool steels. *Vacuum* **2017**, *135*, 7–12. [CrossRef]
9. Zhang, S.; Zhu, W. TiN coating of tool steels: A review. *J. Mater. Process. Technol.* **1993**, *39*, 165–177. [CrossRef]
10. Hsu, C.H.; Lin, C.Y.; Chen, J.X. Wear and corrosion performance of Ti-6Al-4V alloy arc-coated TiN/CrN nano-multilayer film. *Metals* **2023**, *13*, 907. [CrossRef]
11. Atar, E.; Alpaslan, Ö.; Çelik, Ö.; Çimenoĝlu, H. Tribological properties of CrN coated H13 grade tool steel. *J. Iron Steel Res. Int.* **2014**, *21*, 240–245. [CrossRef]
12. Zhang, C.; Gu, L.; Tang, G.; Mao, Y. Wear transition of CrN coated M50 steel under high temperature and heavy load. *Coatings* **2017**, *7*, 202. [CrossRef]
13. Singh, A.; Kumar, N.; Kuppusami, P.; Prasanthi, T.N.; Chandramohan, P.; Dash, S.; Srinivasan, M.P.; Mohandas, E.; Tyagi, A.K. Tribological properties of sputter deposited ZrN coatings on titanium modified austenitic stainless steel. *Wear* **2012**, *280–281*, 22–27. [CrossRef]
14. Tao, H.; Zhylinski, V.; Vereschaka, A.; Chayeuski, V.; Yuanming, H.; Milovich, F.; Sotova, C.; Seleznev, A.; Salychits, O. Comparison of the mechanical properties and corrosion resistance of the Cr-CrN, Ti-TiN, Zr-ZrN, and Mo-MoN coatings. *Coatings* **2023**, *13*, 750. [CrossRef]
15. Bobzin, K.; Kalscheuer, C.; Carlet, M.; Tayyab, M. Influence of aluminum content on the impact fatigue of HPPMS CrAlN coatings on tool steel. *Phys. Mesomech.* **2021**, *24*, 625–632. [CrossRef]
16. Zhang, X.; Zhao, Y.; Gao, W.; Ren, L.; Yang, K. Study of TiCuN/ZrN multilayer coatings with adjustable combination properties deposited on TiCu alloy. *Vacuum* **2022**, *202*, 111202. [CrossRef]
17. Hsu, C.H.; Chen, M.L.; Lai, K.L. Corrosion resistance of TiN/TiAlN-coated ADI by cathodic arc deposition. *Mater. Sci. Eng. A* **2006**, *421*, 182–190. [CrossRef]
18. Kumar, S.; Maity, S.R.; Patnaik, L. Morphology and wear behavior of monolayer TiAlN and composite AlCrN/TiAlN-coated plasma-nitrided DAC-10 tool steel. *J. Sci. Eng.* **2022**, *47*, 15519–15538. [CrossRef]
19. Hernández-Sierra, M.T.; Aguilera-Camacho, L.D.; Ponce, A.; García-Miranda, J.S.; Moreno, K.J. Tribological performance of TiN and TiCN coatings on a working tool steel. *J. Mech. Sci. Technol.* **2018**, *32*, 3659–3666. [CrossRef]
20. Özkan, D.; Yilmaz, M.A.; Karakurt, D.; Szala, M.; Walczak, M.; Bakdemir, S.A.; Türküz, C.; Sulukan, E. Effect of AISI H13 steel substrate nitriding on AlCrN, ZrN, TiSiN, and TiCrN multilayer PVD coatings wear and friction behaviors at a different temperature level. *Materials* **2023**, *16*, 1594. [CrossRef]
21. Kumar, S.; Maity, S.R.; Patnaik, L. Effect of heat treatment and TiN coating on AISI O1 cold work tool steel. *Mater. Today Proc.* **2020**, *26 Pt 2*, 685–688. [CrossRef]
22. Kim, S.K.; Le, V.V. Cathodic arc plasma deposition of nano-multilayered ZrN/AlSiN thin films. *Surf. Coat. Technol.* **2011**, *206*, 1507–1510. [CrossRef]
23. Illana, A.; Almandoz, E.; Fuentes, G.G.; Pérez, F.J.; Mato, S. Comparative study of CrAlSiN monolayer and CrN/AlSiN superlattice multilayer coatings: Behavior at high temperature in steam atmosphere. *J. Alloys Compd.* **2019**, *778*, 652–661. [CrossRef]
24. Bakalova, T.; Petkov, N.; Bahchedzhiev, H.; Kejzlar, P.; Louda, P.; Ďurák, M. Improving the tribological and mechanical properties of an aluminum alloy by deposition of AlSiN and AlCrSiN coatings. *Manuf. Technol.* **2017**, *17*, 824–830.
25. Hsu, C.H.; Huang, W.C.; Lee, Y.P.; Ho, W.Y. Effect of nitrogen atmosphere heat treatment on structure and wear behavior of CrAlSiN nanocomposite film. *Surf. Coat. Technol.* **2017**, *320*, 230–234. [CrossRef]
26. Chang, Y.Y.; Lai, H.M. Wear behavior and cutting performance of CrAlSiN and TiAlSiN hard coatings on cemented carbide cutting tools for Ti alloys. *Surf. Coat. Technol.* **2014**, *259*, 152–158. [CrossRef]
27. Chang, Y.Y.; Chiu, W.T.; Hung, J.P. Mechanical properties and high temperature oxidation of CrAlSiN/TiVN hard coatings synthesized by cathodic arc evaporation. *Surf. Coat. Technol.* **2016**, *303*, 18–24. [CrossRef]
28. Chen, T.K.; Shun, T.T.; Yeh, J.W.; Wong, M.S. Nanostructured nitride films of multi-element high-entropy alloys by reactive DC sputtering. *Surf. Coat. Technol.* **2004**, *189*, 193–200. [CrossRef]

29. Kuczyk, M.; Krülle, T.; Zawischa, M.; Kaspar, J.; Zimmer, O.; Leonhardt, M.; Leyens, C.; Zimmermann, M. Microstructure and mechanical properties of high entropy alloy nitride coatings deposited via direct current cathodic vacuum arc deposition. *Surf. Coat. Technol.* **2022**, *448*, 128916. [CrossRef]
30. Kovalev, A.; Wainstein, D.; Konovalov, E.; Vakhrushev, V.; Fox-Rabinovich, G.; Fox-Rabinovich, M.; Dmitrievskii, S.; Tomchuk, A. Features of tribooxidation of the high-entropy coating (AlCrZrTiTa)N during dry high-speed cutting. *Coatings* **2023**, *13*, 1508. [CrossRef]
31. Heinke, W.; Leyland, A.; Matthews, A.; Berg, G.; Friedrich, C.; Broszeit, E. Evaluation of PVD nitride coatings, using impact, scratch and Rockwell-C adhesion tests. *Thin Solid Films* **1995**, *270*, 431–438. [CrossRef]
32. Oliver, W.C.; McHargue, C.J. Characterizing the hardness and modulus of thin films using a mechanical properties microprobe. *Thin Solid Films* **1988**, *161*, 117–122. [CrossRef]
33. Hsu, C.H.; Lin, C.Y.; You, W.S. Microstructure and dry/wet tribological behaviors of 1% Cu-alloyed austempered ductile iron. *Materials* **2023**, *16*, 2284. [CrossRef] [PubMed]
34. Gonzalez-Carmona, J.M.; Mambuscay, C.L.; Ortega-Portilla, C.; Hurtado-Macias, A.; Jeferson Fernando Piamba, J.F. TiNbN Hard coating deposited at varied substrate temperature by cathodic arc: Tribological performance under simulated cutting conditions. *Materials* **2023**, *16*, 4531. [CrossRef] [PubMed]
35. Lin, J.; Sproul, W.D.; Moore, J.J.; Lee, S.; Myers, S. High rate deposition of thick CrN and Cr2N coatings using modulated pulse power (MPP) magnetron sputtering. *Surf. Coat. Technol.* **2011**, *205*, 3226–3234. [CrossRef]
36. Hsu, C.H.; Huang, K.H.; Lin, Y.H. Microstructure and wear performance of arc-deposited Ti-N-O coatings on AISI 304 stainless steel. *Wear* **2013**, *306*, 97–102. [CrossRef]
37. Leyland, A.; Matthews, A. On the significance of the H/E ratio in wear control: A nanocomposite coating approach to optimized tribological behavior. *Wear* **2000**, *246*, 1–11. [CrossRef]

materials

MDPI

Article

Zirconia and Crofer Joint Made by Reactive Air Brazing Using the Silver Base Paste and Cu-Ti Coating Layer

Shu-Wei Chang [1,2], Ren-Kae Shiue [1,*] and Liang-Wei Huang [3]

[1] Department of Materials Science and Engineering, National Taiwan University, Taipei 106, Taiwan; d03527009@ntu.edu.tw
[2] Chemistry and Environment Research Laboratory, Taiwan Power Research Institute, Taiwan Power Company, New Taipei City 238, Taiwan
[3] Department of Material Research, National Atomic Research Institute, Taoyuan 325, Taiwan; i13501350@nari.org.tw
* Correspondence: rkshiue@ntu.edu.tw

Abstract: This study proposes a method to enhance the airtightness of the joint between the ZrO_2 and Crofer alloy using coating technology. With the aid of vacuum sputtering technology, a titanium–copper alloy layer with a thickness between 1.5 μm and 6 μm was first deposited on the surface of ZrO_2 and Crofer, respectively. The chemical composition of the deposited reaction layer was 70.2 Cu and 29.8 Ti in at%. Then, using silver as the base material in the reactive air brazing (RAB) process, we explore the use of this material design to improve the microstructure and reaction mechanism of the joint surface between ceramics and metal, compare the effects of different pretreatment thicknesses on the microstructure, and evaluate its effectiveness through air tightness tests. The results show that a coating of Cu-Ti alloy on the ZrO_2 substrate can significantly improve bonding between the Ag filler and ZrO_2. The Cu-Ti metallization layer on the ZrO_2 substrate is beneficial to the RAB. After the brazing process, the coated Cu-Ti layers form suitable reaction interfaces between the filler, the metal, the filler, and the ceramic. In terms of coating layer thickness, the optimized 3 μm coated Cu-Ti alloy layer is achieved from the experiment. Melting and dissolving the Cu-Ti coated layer into the ZrO_2 substrate results in a defect-free interface between the Ag-rich braze and the ZrO_2. The air tightness test result shows no leakage under 2 psig at room temperature for 28 h. The pressure condition can still be maintained even under high-temperature conditions of 600 °C for 24 h.

Keywords: reactive air brazing; interface; metallization; microstructure; airtight

1. Introduction

Metals and ceramics are used in different industrial fields due to their specific characteristics. However, in some cases, when taking advantage of two materials at the same time, the two materials must be combined by joining, such as gas turbine blade coatings for power plants [1,2], circuit boards [3,4], biomedical applications [5,6], and fuel cells [7]. Currently known joining technologies for ceramics and metal include brazing, diffusion bonding, and friction welding [8]. Dissimilar metal and ceramic brazing is the primary approach to joining. The braze melt wets the ceramic's surface, forming a metallurgical bond between the ceramic and metal [9].

The brazing technology can provide a joint with relatively good gas tightness between the ceramic and metal. It can also offer better thermal cycle stability, contributing to the joint's long-term reliability. It is a sealing approach with potential for industrial applications [10–12]. AMB (active metal brazing) and RAB (reactive air brazing) are well-known joining technologies for metal-ceramic sealing. AMB uses an inert gas, such as Ar and He, or a relatively high-vacuum ($<10^{-8}$ bar) in brazing to avoid the deterioration of the metallic filler alloy. Higher production costs are necessary. Moreover, the low oxygen partial pressure needed in the brazing temperature may lead to irreversible degradation of the ceramic substrate [13]. In contrast, RAB is a potential alternative to producing an airtight metal/ceramic joint. It has a lower manufacturing cost without degradation of the ceramic substrate.

ZrO_2 has been widely applied in SOFCs (solid oxide fuel cells), TBC (Thermal Barrier Coating), and the electronic industry since it is featured with high thermal and chemical stability [7,14,15]. Due to the similar thermal expansion coefficient to ZrO_2, the Crofer is considered a suitable alloy to join with ZrO_2. In recent years, a few scholars have proposed using silver as the sealing metal, introducing a few percent oxide to improve its coefficient of thermal expansion mismatch between the Ag and ZrO_2 [16–18]. Some scholars also use low-melting point commercial active solders to join the metal and ceramics [19,20]. It has also been reported that coating technology is used to metalize the surface of ceramics to improve their wettability, allowing brazing to proceed [21,22].

The commercially available Ticusil active braze alloy, 26.7 Cu, 4.5 Ti, and 68.8 Ag in wt%, is a popular filler metal in brazing ZrO_2 and stainless steel [19,23,24]. However, the oxidation of such a brazed joint deteriorates rapidly because Cu is particularly susceptible to oxidation in high-temperature air environments [25]. Decreasing the Cu amount in the Ag-based braze could be a possible solution to overcome the above issue. Laik et al. [26] used Ag-based commercial active braze alloy to vacuum braze Al_2O_3/Inconel 600, which was stable and kept gas-tight and robust. Wang et al. [27] use Ag-5Cu-1Al-1.25Ti (wt%) filler to vacuum braze Al_2O_3 and Kovar 4 J34 alloy. Based on these research findings, the feasibility of applying Ag-based filler alloyed with Ti and Cu as brazing material was evidenced to achieve a solid joint for ceramic and metal. Suppose the copper content of the braze alloy can be further reduced through the design of ceramic metallization before the brazing process. In that case, the central issue of silver-based filler may be solved.

Thin film sputtering coating(s) can be recognized as one of the metallization methods used for ceramic pretreatment before brazing. In a previous study, pure Cu, Ti, and Ag layers were sequentially deposited on the ZrO_2 substrate, and vacuum brazing of the metalized ZrO_2 and Crofer alloy using the 72Ag-28Cu (wt%) filler foil was performed [20]. However, the melting points of Cu (1084.87 °C) and Ti (1670 °C) are much higher than that of 70Cu-30Ti in at%, approximately 875 °C [28]. The deposition of the 70Cu-30Ti alloy coating layer on the ZrO_2 substrate is expected to provide a higher activity in brazing ZrO_2 and Crofer [19,29,30].

This study proposes a thin film sputtering 70Cu-30Ti (at%) alloy coating layer for ZrO_2 pretreatment before brazing. Then, it uses the silver paste as the braze filler for the RAB process. This study also discusses how the coating layer design improves the microstructure and reaction mechanism of the ceramic-metal joint and compares different pretreatment thicknesses of the Cu-Ti coating layer on ZrO_2. Based on the airtightness of the RAB zirconia and Crofer joint, an optimized thickness of the 70Cu-30Ti coating layer is achieved in this study.

2. Materials and Experimental Procedures

2.1. Materials Preparation

The ZrO_2 plate was purchased from Kceracell, Geumsan-gun, Republic of Korea. The Crofer plate was purchased from VDM Metals GmbH, Altena, Germany. The chemical composition in wt% is 23 Cr, 0.03 C, 0.8 Mn, 0.2 Si, 0.1 Al, 2.0 W, 0.8 Nb, 0.5 Ni, 0.5 Cu, and Fe balance. ZrO_2 and Crofer substrates were cut into specimens with $10 \times 10 \times 1$ mm^3 in dimension. PELCO® colloidal silver was purchased from TED PELLA, Redding, CA, USA. PELCO® colloidal silver with 60 ± 1 wt% Ag, 10~25 wt% toluen, 5~10 wt% ethyl acetate, 5~10wt% ethanol, 5~10 wt% n-butyl acetate, <2.5 wt% propan-2-ol, and <2.5 wt% bornan-2-on.

ZrO_2 and Crofer substrates were metalized using a sputtering machine (KaoDuen KD-SPA01, New Taipei City, Taiwan). They were preheated to 200 °C under a vacuum of 6×10^{-6} mbar, and then argon was introduced to reach a vacuum of about 3×10^{-2} mbar. The temperature of the rotating disk was controlled at 200 °C. In the double target Ti

(170 W) and Cu (80 W) sputtering machine, the thicknesses of the coating films (1.5 μm, 3 μm, and 6 μm) were controlled by different sputtering periods, 1 h, 2 h, and 4 h. The chemical composition of the coated Cu-Ti alloy layer was 70Cu and 30Ti in at%. An electron microscope (Hitachi S-4800, Tokyo, Japan) and energy dispersive spectrometer (Bruker-Quantax, Berlin, Germany) were used to analyze the chemical composition of the coated layer with 70.2 Cu and 29.8 Ti in at%.

2.2. RAB Process

In the RAB, the brazing was performed in a furnace (Ultra Fine UF-E1FS, New Taipei City, Taiwan) at 960 °C with a heating rate of 60 °C/h, held for 20 min, and cooled to 25 °C with a 60 °C/h cooling rate in the air. The cross-section of the brazing specimen was cut with the aid of a slow-speed cutter. The specimen was hot-mounted using the conductive epoxy resin, ground, and polished using a standard metallographic procedure. The sizes of the diamond polishing particles were 9 μm, 3 μm, and 1 μm, respectively. An electron microscope (Hitachi S-4800) and an energy dispersive spectrometer (Bruker-Quantax) were used to analyze the microstructure and phase compositions of the brazed joint. Quantitative chemical analysis was performed using a field emission electron probe microanalyzer (JEOL JXA-8530F Plus, Tokyo, Japan) with a minimum spot size of 1 μm.

2.3. Pressure Drop Test

To evaluate the sealing performance of the brazed specimen, a pressure drop method was used to measure the airtightness of the sealed sample [20,21]. Figure 1 shows the schematic diagram of the RAB specimen used in the pressure drop test. The Crofer alloy and ZrO_2 are 14 × 14 × 1.5 mm^3 and 20 × 20 × 1 mm^3. The Crofer substrate has drilled a hole with a diameter of 8 mm in the central area of the Crofer alloy in advance and connected to the gas source through a metal tube, as displayed in Figure 1a. The flow gas was controlled by a ball valve, as shown in Figure 1b. The internal pressure of the brazed specimen to be tested was recorded over time to calculate the gas leakage rate.

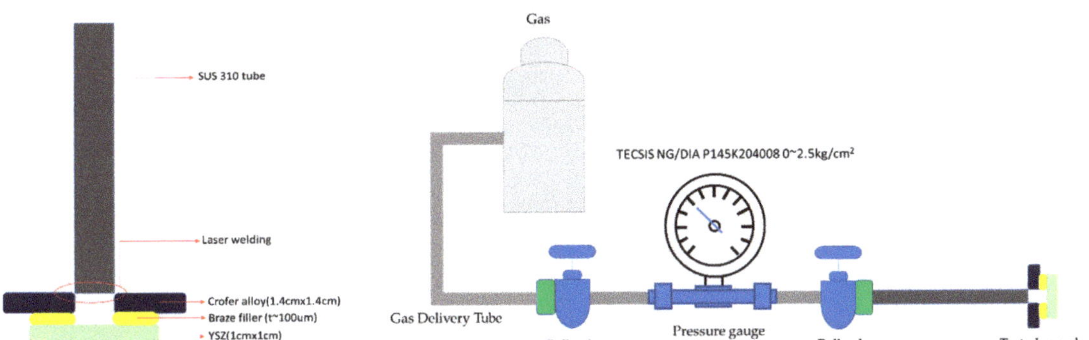

Figure 1. The schematic diagram of the pressure drop test: (**a**) the brazed specimen, (**b**) test infrastructure.

3. Results and Discussion

For comparison purposes, EPMA microstructural observations and quantitative element analyses were carried out on the cross-section of samples without the titanium–copper coating layer. Both secondary electron image (SEI) and backscattered electron image (BEI) are shown in Figure 2a,b. In Figure 2c,d, interfacial separations were observed among the Ag-rich braze alloy and two substrates, Crofer and ZrO_2. Table 1 shows WDS quantitative chemical analyses of selected locations of the brazed joint without the Cu-Ti coating layer in Figure 2c,d. In Figure 2c, the Ag-rich matrix and the Crofer substrate were observed, as marked by C and A. According to Table 1, there was a layer of chromium oxide, as marked by B in Figure 2c, and the thickness of the chromium oxide layer on the interface was approximately 1 μm. The stoichiometric ratio between Cr and O at position B was close to Cr_2O_3. Figure 2d shows the microstructural observation of the interface between the Ag-rich matrix and the ZrO_2 substrate. Figure 2d shows an apparent interfacial crack between the ZrO_2 substrate and the Ag-rich braze. The Ag filler cannot be applied in RAB ZrO_2 and Crofer alloy due to oxidation of the Crofer substrate in the air and high residual stresses resulting from the mismatch of thermal expansion coefficients between the Ag filler metal and ZrO_2. An inferior brazed joint with interfacial separations was observed among the Ag filler metal, ZrO_2, and Crofer substrate.

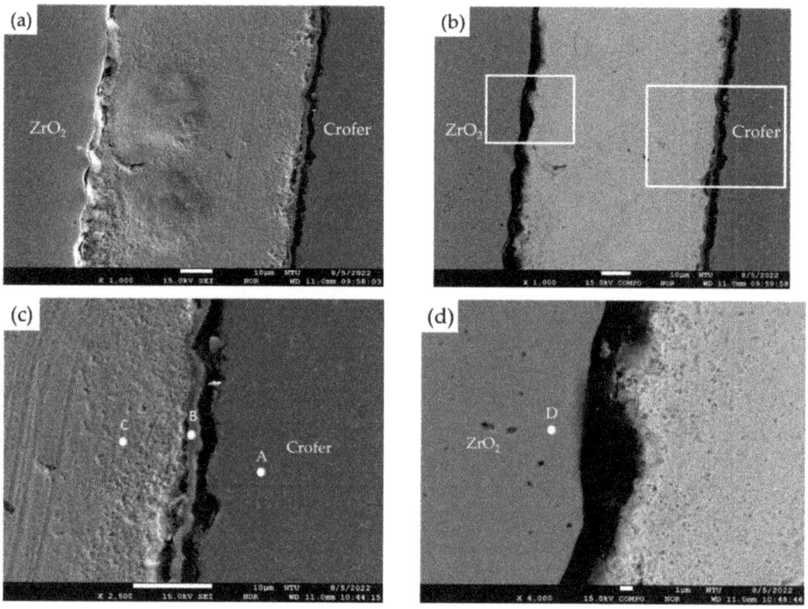

Figure 2. (**a**) SEI, (**b**) BEI of the ZrO_2/Ag/Crofer joint reactive air brazed at 960 °C for 1200 s; (**c**) higher magnification at the Ag/Crofer interface, (**d**) higher magnification at the ZrO_2/Ag interface.

Table 1. EPMA quantitative chemical analyses of the reactive air-brazed joint without coating in Figure 2c,d.

Element/at%	Ag	Cu	Cr	Fe	O	Ti	Y	Zr	Alloy/Phase
A	0.0	0.0	23.9	73.8	0.00	0.0	0.0	0.0	Crofer (substrate)
B	1.2	0.0	35.0	2.1	59.2	0.0	0.0	0.0	Cr_2O_3
C	97.6	0.0	0.5	1.0	0.5	0.0	0.0	0.0	Ag-rich
D	0.0	0.0	0.0	0.0	63.3	0.0	2.0	34.6	ZrO_2 (substrate)

For the RAB joint with a titanium–copper coating layer of 3 μm in thickness on the ZrO_2 substrate, EPMA microscopic observation and quantitative chemical analyses were carried out, as illustrated in Figure 3 and Table 2. The SEI and BEI of the Crofer/Cu-Ti/ZrO_2 joint are shown in Figure 3a,b, respectively. The interfaces among the Ag-rich braze and two substrates were significantly improved compared to those in Figure 2a,b, and no separation was observed at the interfaces. Strong interfacial reactions of the coated Cu-Ti layer and ZrO_2 substrate were observed in Figure 3d. Figure 3c shows the EPMA quantitative chemical analyses of selected positions at the Ag/Crofer interface, and the results are shown in Table 2. In Figure 3c, these oxides primarily alloyed with Cu, Cr, and Fe at positions E, F, and H. TiO_2 (marked by G) and Fe/Ti-rich oxide (marked by J) were also observed with a darker color. The interfacial reaction layer was mainly composed of oxide, and a few Ag-rich phases (marked by I and K) were dispersed in the oxide layer. The reactive wetting formed a good bond between the coated Cu-Ti layer and ZrO_2. In Figure 3d, there was an interfacial layer between the Ag-rich matrix (marked by Q and R) and the ZrO_2 substrate (marked by P), and a Ti-rich oxide was observed as marked by M. The interfacial layer comprised ZrO_2 alloyed with different Ti/Cu contents, as marked by N and O in Table 2. Dissolving the Cu-Ti coated layer into the ZrO_2 substrate resulted in a defect-free interface between the Ag-rich braze and substrate, as demonstrated by Figure 3a–d.

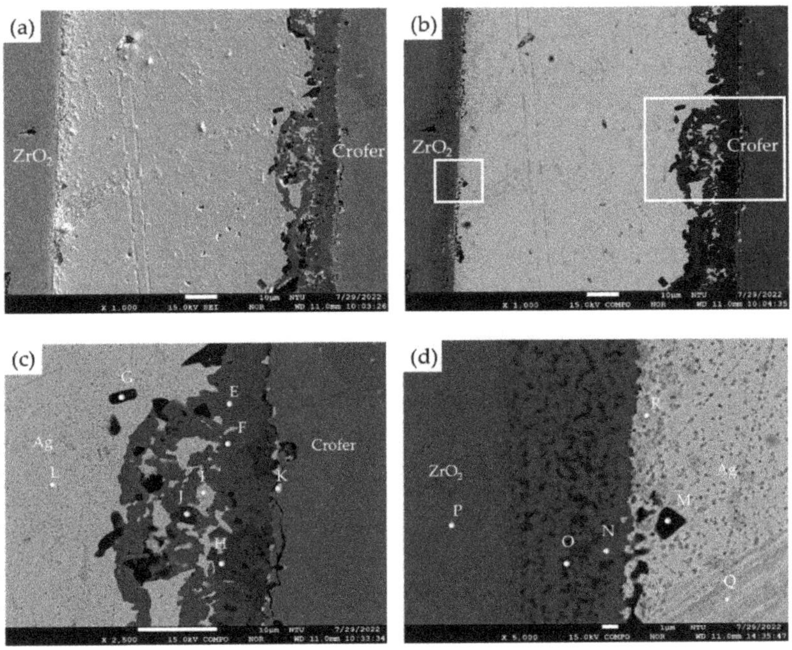

Figure 3. (**a**) SEI, (**b**) BEI of the ZrO_2/Ag/Crofer joint reactive air brazed with 3 μm coating layer at 960 °C for 1200 s; (**c**) higher magnification at the Ag/Crofer interface, (**d**) higher magnification at the ZrO_2/Ag interface.

For the RAB joint without the 70Cu-30Ti active coating layer, interfacial separations were extensively observed at the interfaces, as displayed in Figure 2. There was no active ingredient in the braze filler or Ag paste. The wettability of the Ag paste was poor for the ZrO_2 substrate. Additionally, the wettability of the Ag paste on the oxidized Crofer substrate was also poor during RAB. In contrast, the Cu-Ti alloy coating layer was the only active ingredient during the RAB in the experiment. It formed a Cu/Ti-rich liquid at 960 °C with high activity for both substrates, ZrO_2 and Crofer [30,31]. The reactive wetting of both substrates was achieved, and the Cu/Ti-rich liquid was reacted and solidified as part of the reaction layers.

The interfacial reaction of the ZrO_2 side differed from that of the Crofer side, as illustrated in Figure 3. According to the liquidus projection of the Ag-Cu-Ti ternary alloy phase diagram, there were two immiscible liquids (Ag-rich and Cu/Ti-rich) above 900 °C [30,31]. The activity of the Ag-rich liquid was much less than that of the Cu/Ti-rich liquid. The Ag-rich liquid acted as a barrier against the oxidation of the braze alloy during the RAB. The ZrO_2 substrate was readily wetted by alloying the Cu/Ti liquid, as demonstrated in Figure 3d and Table 2. For the Crofer side, the dissolution of Cr and Fe from the substrate into the Cu/Ti-rich melt and subsequently oxidation during

RAB resulted in islands of Cr/Cu/(Fe)-rich oxides away from the interface, instead of an intermediate layer forming directly at the interface, as shown in Figure 3c and Table 2. The Cr-O binary alloy phase diagram indicates that Cr_2O_3 is the only oxide [28]. The formation of Cr_2O_3 in Figure 2c without the 70Cu-30Ti coating layer was detrimental to the bonding between the Ag filler and Crofer substrate. Separation of the Cr_2O_3 and Crofer substrate was observed in Figure 2c. In contrast, Cr/Cu/(Fe)-rich oxides formed in the interface with 3 μm 70Cu-30Ti active coating layer after RAB was free of crack, as shown in Figure 3c.

Table 2. EPMA quantitative chemical analyses of the reactive air-brazed joint with a 3 μm Cu-Ti coating layer in Figure 3c,d.

Element/at%	Ag	Cu	Cr	Fe	O	Ti	Y	Zr	Alloy/Phase
Crofer	0.0	0.1	16.8	81.2	0.0	0.0	0.0	0.0	Crofer substrate
E	1.7	23.5	19.6	4.2	49.7	0.8	0.0	0.0	Cr/Cu-rich oxide
F	0.4	14.3	13.9	11.1	57.0	1.4	0.0	0.0	Cr/Cu/Fe-rich oxide
G	2.2	0.0	0.6	0.5	64.5	31.3	0.0	0.0	TiO_2
H	0.2	13.3	12.7	13.3	57.8	1.5	0.0	0.0	Cr/Cu/Fe-rich oxide
I	94.5	2.2	0.8	0.9	1.2	0.1	0.0	0.0	Ag-rich
J	0.5	4.5	1.7	16.7	57.6	18.5	0.0	0.0	Fe/Ti-rich Oxide
K	81.2	2.7	4.6	8.4	2.2	0.1	0.0	0.0	Ag-rich
L	99.6	0.0	0.0	0.0	0.1	0.1	0.0	0.0	Ag-rich
M	3.4	14.6	0.2	0.0	58.4	20.3	2.5	0.4	Ti-rich oxide
N	0.1	0.3	0.0	0.0	63.8	3.1	0.6	32.1	ZrO_2 alloyed with Ti
O	0.1	2.7	0.2	0.0	62.5	8.5	3.5	21.8	ZrO_2 alloyed with Ti/Cu
P	0.0	0.0	0.0	0.0	64.3	0.1	1.9	33.7	ZrO_2 substrate
Q	99.6	0.0	0.0	0.0	0.2	0.0	0.0	0.0	Ag-rich
R	93.3	0.2	0.0	0.1	1.1	0.2	0.0	4.5	Ag-rich

The oxidation of the interface for the Crofer side was much more vigorous than that of the ZrO_2 side. The thickness of the interfacial reaction layer on the Crofer side was approximately between 10 and 20 μm. In contrast, the interfacial thickness of the ZrO_2 side was approximately 5 μm.

Figure 4 shows the EPMA element mappings of the Crofer side in Figure 3c. The interfacial oxide layer mainly comprises Cr, Cu, and Fe. The Ti-rich oxide particles are dispersed in the oxide layer. The Cr element is enriched at the interface close to the Crofer substrate, and oxidation of the Crofer substrate is observed after RAB. The EPMA element mappings at the interface between the Ag braze and ZrO_2 substrate in Figure 3d are shown in Figure 5. Unlike the interface of the Crofer side displayed in Figure 4, the distribution of Cu, O, and Ti agrees with the quantitative chemical analysis results shown in Table 2. The Cu/Ti-rich melt strongly reacts with the ZrO_2 substrate and promotes reactive wetting of the ZrO_2 substrate [19]. Both EPMA mappings are consistent with the aforementioned quantitative chemical analysis results. The Cu-Ti metallization layer on the ZrO_2 substrate is beneficial to the RAB.

To evaluate the effect of Cu-Ti coated layer thickness on the RAB joints, RAB was conducted with different Cu-Ti coating thicknesses. Both SEI and BEI of the RAB specimen with a 1.5 μm Cu-Ti layer thickness are shown in Figure 6. Under the condition of a 1.5 μm Cu-Ti coating layer, the thickness of the interfacial reaction layer between the Ag-rich braze and ZrO_2 is significantly decreased to approximately 5 μm, as displayed in Figure 6d. However, there are many pores and cracks, as shown in Figure 6a–c. It is expected that the RAB joint with a 1.5 μm Cu-Ti coating layer on the substrate is not gas-tight. The microstructure of the interface between the Ag braze and ZrO_2 is shown in Figure 6d. Similarly, the thickness of the interface in Figure 6d is much less than that of 3d.

Table 3 shows EPMA WDS quantitative chemical analyses of selected positions in the RAB joint with a 1.5 μm Cu-Ti coated layer in Figures 6c and 6d, respectively. The 1.5 μm Cu-Ti coated layer on the Crofer substrate was alloyed with Cr and Fe, and was oxidized into Cr/Cu/(Fe)-rich oxide as marked by S, T, U, and V in Figure 6c. For the ZrO_2 side of the RAB joint, the ZrO_2 substrate close to the interface was alloyed with Ti and Cu, as marked by W, X, and Y in Figure 6d. Cu and Ti concentrations decreased with increasing distance from the braze/ZrO_2 interface.

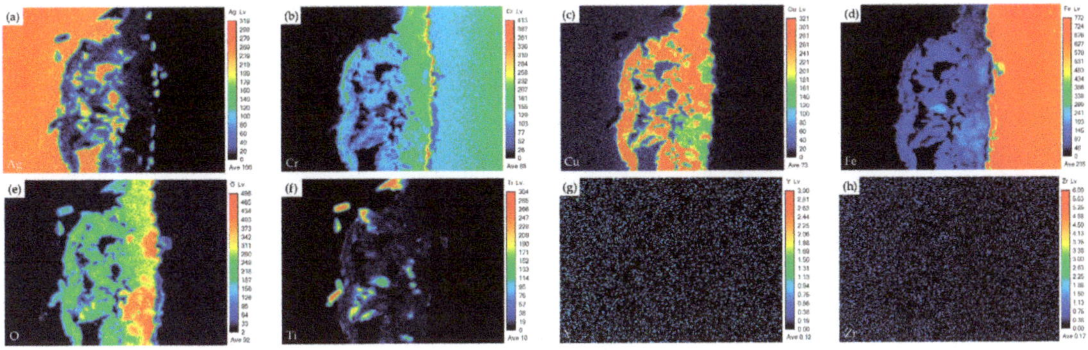

Figure 4. EPMA element mappings of the Crofer side interface in Figure 3c: (**a**) Ag, (**b**) Cr, (**c**) Cu, (**d**) Fe, (**e**) O, (**f**) Ti, (**g**) Y, and (**h**) Zr.

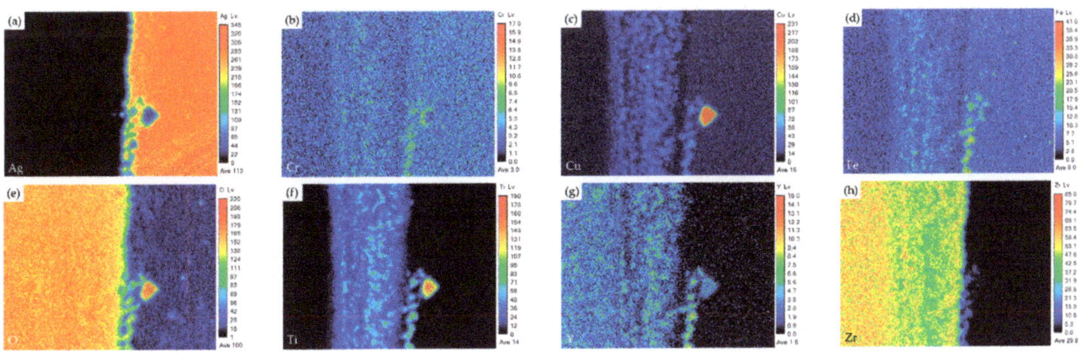

Figure 5. EPMA element mappings of the ZrO_2 side interface in Figure 3d: (**a**) Ag, (**b**) Cr, (**c**) Cu, (**d**) Fe, (**e**) O, (**f**) Ti, (**g**) Y, and (**h**) Zr.

Figure 7 shows EPMA element mappings of the Crofer side in Figure 6c. The interfacial oxide layer mainly comprises Cr, Cu, and Fe. Less Ti-rich oxide particles are observed in the oxide layer. The Cr element is enriched at the interface close to the Crofer substrate, and oxidation of the Crofer substrate is observed after RAB. The EPMA element mappings at the interface between the Ag-rich braze and ZrO_2 substrate (Figure 6d) are shown in Figure 8. It has been reported that The Cu/Ti-rich melt has a strong reaction with the ZrO_2 substrate and promotes reactive wetting of the ZrO_2 substrate. Both EPMA mappings are consistent with the aforementioned quantitative chemical analysis results.

Strong interfacial reaction and oxidation were observed in the RAB with the 1.5 μm Cu-Ti coating layer. However, thinner reaction/oxidation interfacial layers were observed compared to those of the 3 μm Cu-Ti coating layer due to less Cu and Ti deposited on the substrate. The RAB joint has many microcracks and pores with the 1.5 μm Cu-Ti coating layer, as illustrated in Figure 6. Because the interfacial reaction/oxidation layers were prone to fracture in the joint with a 1.5 μm Cu-Ti coating layer, the airtightness of the RAB joint is inferior to that of the joint with a 3 μm Cu-Ti coating layer.

Figure 9 shows an RAB joint's SEI and BEI images with a 6 μm Cu-Ti coating layer thickness. The thickness of the interfacial reaction layer on the Crofer side was significantly increased to approximately 30 μm, as illustrated in Figure 9a,b. The Cu-Ti alloy coating formed a Cu/Ti-rich liquid with high activity for both substrates, ZrO_2 and Crofer. The

Cu/Ti-rich liquid was reacted and oxidized as part of the reaction layers next to the substrates, as displayed in Figure 9. It was noted that pores were in the interfacial reaction layer, as indicated by arrows in Figure 9a. The maximum size of the pore was approximately 10 μm.

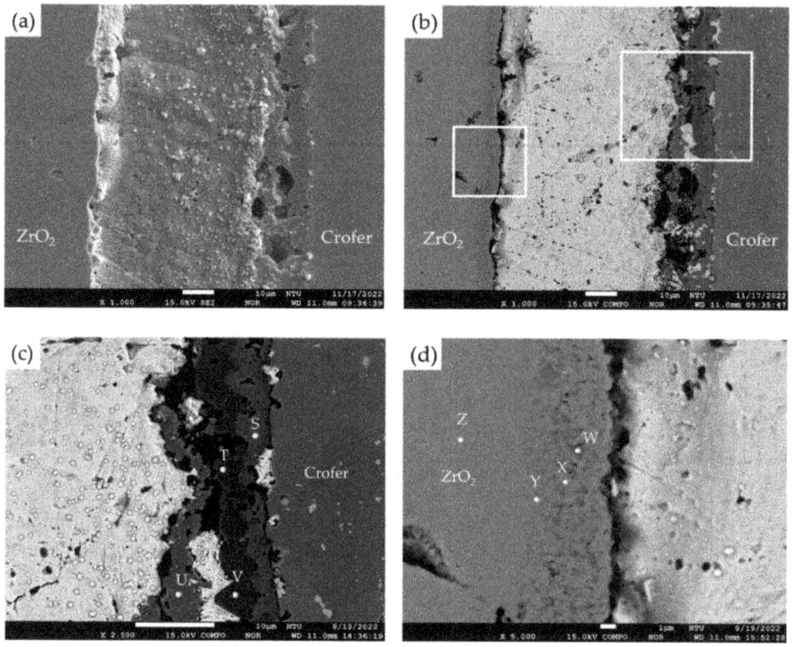

Figure 6. (**a**) SEI, (**b**) BEI of the ZrO_2/Ag/Crofer joint brazed with 1.5 μm coated film at 960 °C for 1200 s, (**c**) higher magnification at the Ag/Crofer interface, (**d**) higher magnification at the ZrO_2/Ag interface.

Table 3. EPMA quantitative chemical analyses of the brazed joint with a 1.5 μm Cu-Ti coated layer in Figure 6c,d.

Element/at%	Ag	Cu	Cr	Fe	O	Ti	Y	Zr	Phase/Alloy
S	6.4	37.9	28.7	5.2	20.5	0.8	0.0	0.0	Cr/Cu-rich Oxide
T	0.3	25.4	22.1	19.7	27.7	2.4	0.0	0.0	Cr/Cu/Fe-rich Oxide
U	8.9	37.4	19.9	10.3	20.9	1.6	0.0	0.0	Cr/Cu/Fe-rich Oxide
V	0.5	25.8	21.4	19.1	27.2	2.9	0.0	0.0	Cr/Cu/Fe-rich Oxide
W	0.1	3.6	0.0	0.7	61.6	10.6	4.8	17.6	ZrO_2 alloyed with Ti/Cu
X	0.1	0.6	0.0	0.1	64.9	3.9	1.3	28.7	ZrO_2 alloyed with Ti
Y	0.1	0.4	0.0	0.3	64.8	2.9	0.9	30.5	ZrO_2 alloyed with Ti
Z	0.0	0.0	0.0	0.0	62.4	0.0	3.2	34.3	ZrO_2 (substrate)

The microstructure and EPMA quantitative chemical analysis results of selected positions at the interface of the Crofer side are shown in Figure 9c and Table 4. Regarding chemical composition, the interfacial layer is mainly composed of oxides. There are different oxide particles in the thick Cr/Cu/(Fe)-rich oxide layer, as marked by A1~E1 in Figure 9c. The microstructure and EPMA quantitative chemical analysis results of the ZrO_2 side are shown in Figure 9d and Table 4. The connection between the Ag-rich braze alloy and ZrO_2 is in good shape with regards to the microstructure. The ZrO_2 substrate close to the interface is alloyed with Ti, as marked by F1 in Figure 9d and Table 4. The thickness of both interfacial reaction layers in the RAB with the 6 μm Cu-Ti coating layer is much greater than in the previous cases.

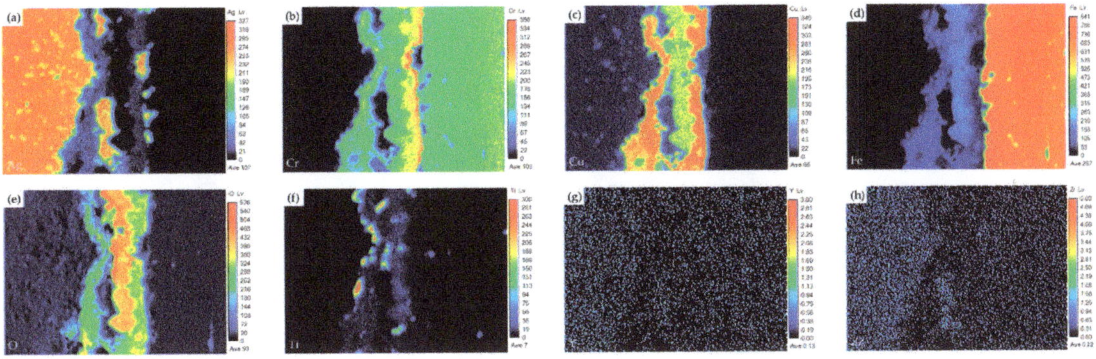

Figure 7. EPMA element mappings of the Crofer side interface in Figure 6c: (**a**) Ag, (**b**) Cr, (**c**) Cu, (**d**) Fe, (**e**) O, (**f**) Ti, (**g**) Y, and (**h**) Zr.

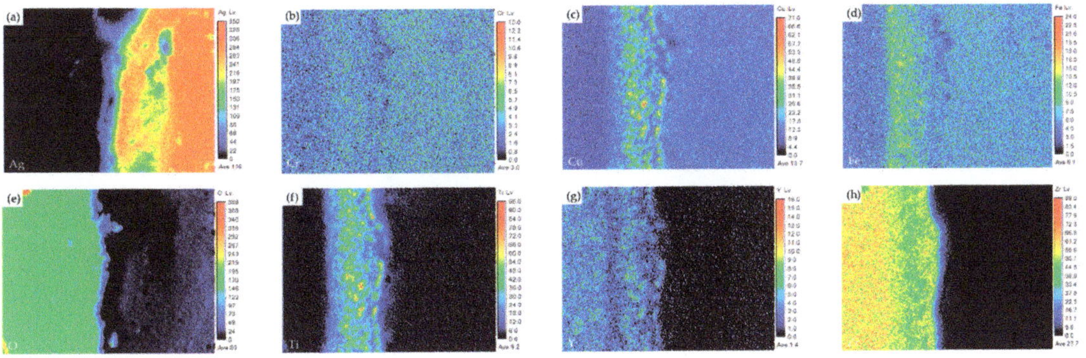

Figure 8. EPMA element mappings of the ZrO_2 side interface in Figure 6d: (**a**) Ag, (**b**) Cr, (**c**) Cu, (**d**) Fe, (**e**) O, (**f**) Ti, (**g**) Y, and (**h**) Zr.

Figure 10 shows the EPMA element mappings of the RAB joint with a 6 μm Cu-Ti coating layer at the interface of the Crofer side. Under the 6 μm film thickness, oxidation of the Cu-Ti coating layer and Crofer substrate is easily identified. Because the Cu-Ti coating film was thicker, the thickness of the interfacial oxide layer at the Crofer side was increased to nearly 30 μm. The interfacial defects at the Crofer side significantly deteriorated the gas-tight performance of the RAB joint. The EPMA element mappings of the interface at the ZrO_2 side are shown in Figure 11. Unlike the Crofer side in Figure 10, the diffusion of copper and titanium elements in the interface layer is slightly different. The concentration of titanium in the interface layer is relatively low. The diffusion depth of titanium and copper elements can reach up to 20 μm, while silver elements do not diffuse into the ceramics.

To evaluate the sealing performance of the brazed specimen, a pressure drop test was used to measure the airtightness of the sealed sample. According to the airtight test method, the 3 μm brazed samples were subjected to airtightness tests at room and high temperatures. The test condition was 2 psig helium and room temperature for 28 h, as shown in Figure 12. The test result shows that the pressure did not drop at room temperature. The same test condition was changed to a high temperature of 600 °C, and the result is shown in Figure 13, which also does not show the pressure drop. Applying a 70Cu-30Ti (at%) alloy coating layer on two substrates before reactive air brazing contributes to excellent airtight performance of the joint.

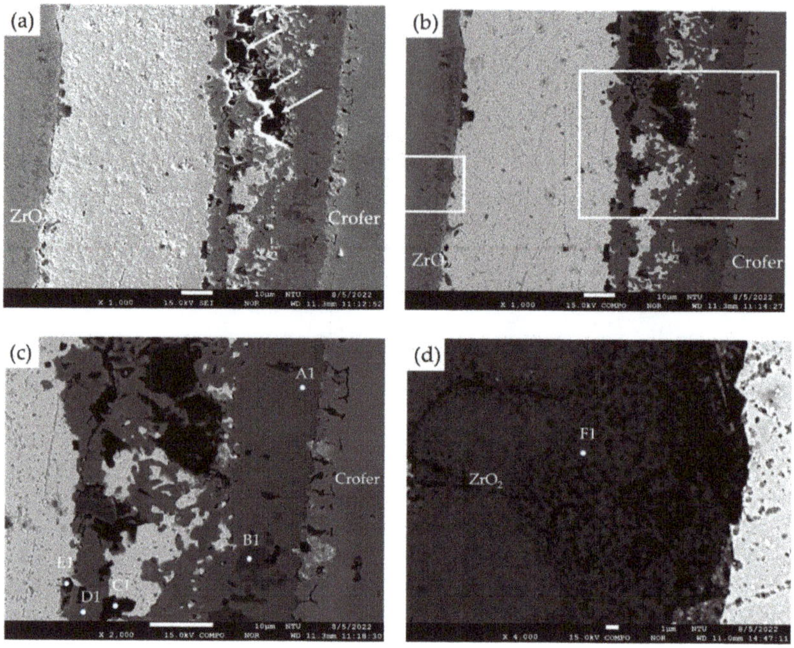

Figure 9. (**a**) SEI, (**b**) BEI of the ZrO_2/Ag/Crofer joint brazed with 6 μm coated film at 960 °C for 1200 s; (**c**) higher magnification at the Ag/Crofer interface, (**d**) higher magnification at the ZrO_2/Ag interface.

Table 4. EPMA quantitative chemical analyses of junction between 6 μm film brazing sample and Crofer in Figure 9.

Element/at%	Ag	Cu	Cr	Fe	O	Ti	Y	Zr	Alloy/Phase
A1	1.9	24.2	19.3	5.9	47.3	0.9	0.0	0.0	Cr/Cu-rich Oxide
B1	0.1	15.3	10.4	15.7	55.5	1.9	0.0	0.0	Cr/Cu/Fe-rich Oxide
C1	0.7	0.3	0.8	1.4	62.4	32.2	0.0	0.0	TiO_2
D1	4.6	22.8	8.3	10.3	50.5	3.2	0.0	0.0	Cr/Cu/Fe-rich Oxide
E1	1.2	0.2	0.6	1.1	62.8	32.4	0.0	0.0	TiO_2
F1	0.0	0.4	0.0	0.0	64.9	3.2	1.0	29.9	ZrO_2 alloyed with Ti

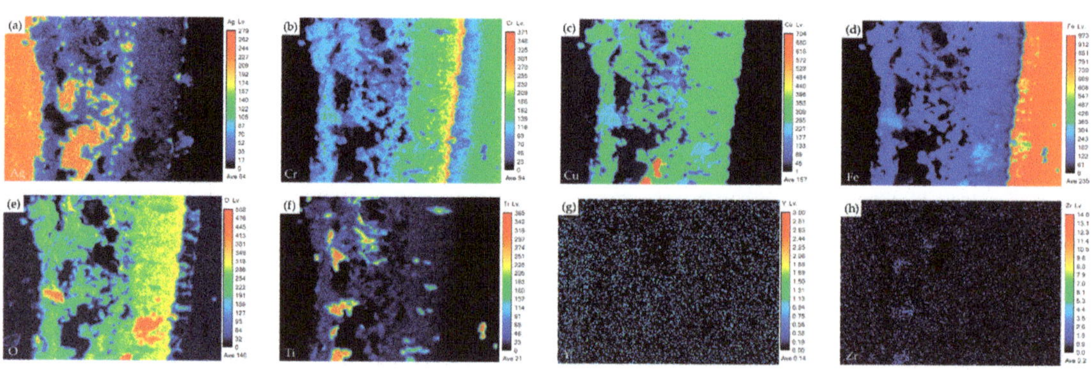

Figure 10. EPMA element mappings of the Crofer side interface in Figure 9c: (**a**) Ag, (**b**) Cr, (**c**) Cu, (**d**) Fe, (**e**) O, (**f**) Ti, (**g**) Y, and (**h**) Zr.

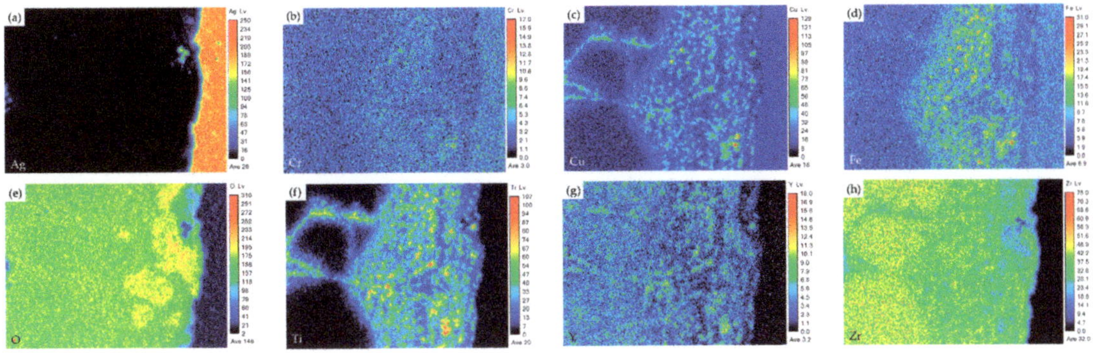

Figure 11. EPMA element mappings of the ZrO_2 side interface in Figure 9d: (**a**) Ag, (**b**) Cr, (**c**) Cu, (**d**) Fe, (**e**) O, (**f**) Ti, (**g**) Y, and (**h**) Zr.

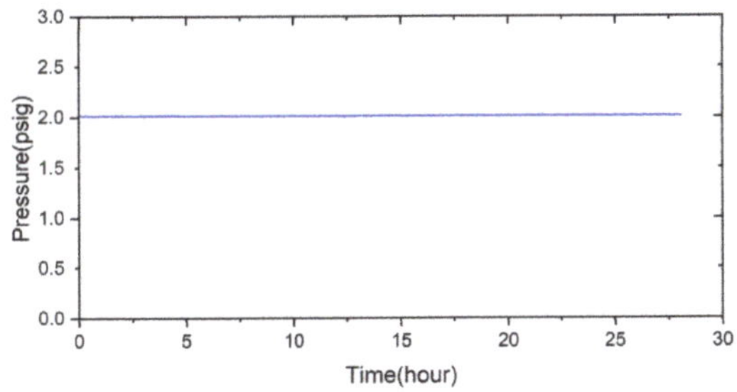

Figure 12. The pressure drop in the joint leak test was performed at room temperature for 28 h.

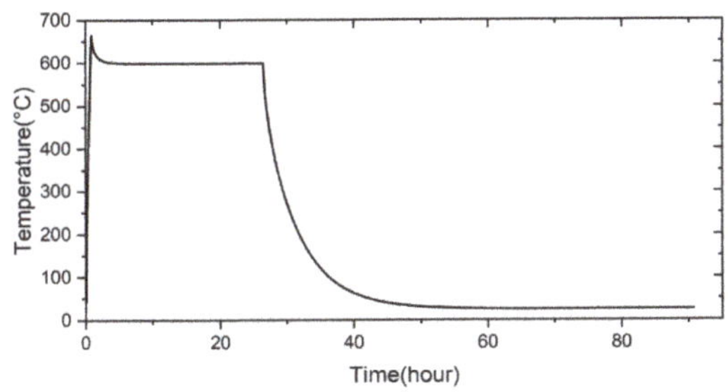

(**a**)

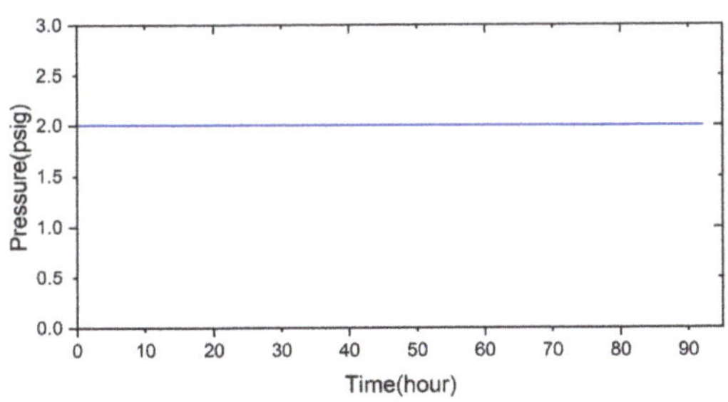

(**b**)

Figure 13. The pressure drop in the joint leak test was at 600 °C for 24 h: (**a**) temperature and (**b**) pressure profiles.

4. Conclusions

A novel method for reactive air brazing at 960 °C for the 1200 s to join Cu-Ti metalized ZrO_2 and Crofer alloy using Ag paste as the braze alloy has been proposed, which proved effective in achieving airtightness of the joint. Important conclusions are summarized below:

1. Applying 70Cu-30Ti (at%) active coating layer on both substrates can effectively improve the bonding between pure silver and ZrO_2 and pure silver and Crofer alloy.
2. The interfacial reaction of the ZrO_2 side is very different from that of the Crofer side. There are two immiscible liquids, Ag-rich and Cu/Ti-rich liquids, at 960 °C. The activity of the Ag-rich liquid is much less than that of the Cu/Ti-rich liquid. The Ag-rich liquid acts as a barrier against the oxidation of the braze alloy during the RAB. The ZrO_2 is readily wetted by alloying the Cu/Ti liquid. For the Crofer side, the Cu/Ti-rich liquid reacted with Fe and Cr in the Crofer substrate and oxidized during the RAB.
3. The oxidation of the interface for the Crofer side was more vigorous than that of the ZrO_2 side. The thickness of the interfacial reaction layer on the Crofer side was approximately 10 and 20 μm. In contrast, the interfacial thickness of the ZrO_2 side was approximately 5 μm.
4. The coating thickness of the 70Cu-30Ti active layer is optimized into 3 μm in the experiment. ZrO_2/Ag and Ag/Crofer interfaces are free of pores and cracks.
5. There is no pressure drop under 2 psig at room temperature for 28 h. The pressure condition can still be maintained even under the high-temperature condition of 600 °C for 24 h. It can potentially be used in solid oxide fuel cells in the future.

Author Contributions: Experiment, S.-W.C. and L.-W.H.; formal analysis, S.-W.C.; writing—original draft preparation, S.-W.C. and R.-K.S.; writing—review and editing, S.-W.C., R.-K.S. and L.-W.H.; funding acquisition, R.-K.S.; Resources, S.-W.C., L.-W.H. and R.-K.S. All authors have read and agreed to the published version of the manuscript.

Funding: This research was funded by the Ministry of Science and Technology, ROC (Contract No. MOST 109-2221-E-002-041-MY3).

Institutional Review Board Statement: Not applicable.

Informed Consent Statement: Not applicable.

Data Availability Statement: The raw data supporting the conclusions of this article will be made available by the authors on request.

Acknowledgments: We thank H.C. Lin and C.Y. Kao of the Instrumentation Center, National Taiwan University, for EPMA and experiments.

Conflicts of Interest: Author Shu-Wei Chang was employed by the Taiwan Power Company. The remaining authors declare that the research was conducted in the absence of any commercial or financial relationships that could be construed as a potential conflict of interest.

References

1. Li, J.; Li, Y. Micro gas turbine: Developments, applications, and critical technologies on components. *Propuls. Power Res.* **2023**, *12*, 1–43. [CrossRef]
2. Moskal, G.; Niemiec, D.; Chmiela, B.; Kałamarz, P.; Durejko, T.; Ziętala, M.; Czujko, T. Microstructural characterization of laser cladding NiCrAlY coatings on Inconel 625 Ni-based superalloy and 316L stainless steel. *Surf. Coat. Technol.* **2020**, *387*, 125317. [CrossRef]
3. Murayama, N.; Hirao, K.; Sando, M.; Tsuchiya, T.; Yamaguchi, H. High-temperature electro-ceramics and their application to SiC power modules. *Ceram. Int.* **2018**, *44*, 3523–3530. [CrossRef]
4. Hirao, K.; Zhou, Y.; Fukushima, M.; Wakasugi, N.; Suganuma, K. Evaluation of in-plane effective thermal conductivity for metalized ceramic substrates using a micro heater chip. *Ceram. Int.* **2023**, *49*, 28593–28606. [CrossRef]
5. Ntasi, A.; Jabbari, Y.S.A.; Mueller, W.D.; Eliades, T.; Zinelis, S. Electrochemical characterization of novel Ag-based brazing alloys for dental applications. *Dent. Mater.* **2019**, *35*, 163–174. [CrossRef]
6. Sun, Q.; Yang, L.; Yang, W.; Ji, H.; Li, M.; Li, Y. Microstructure evolution and bonding mechanism of ZrO_2 ceramic and Ti-6Al-4V alloy joints brazed by Bi_2O_3-B_2O_3-ZnO glass paste. *J. Eur. Ceram. Soc.* **2022**, *42*, 5953–5963. [CrossRef]

7. Hirata, Y.; Daio, S.; Kai, A.; Shimonosono, T.; Yano, R.; Sameshima, S.; Yamaji, K. Performance of yttria-stabilized zirconia fuel cell using H_2-CO_2 gas system and CO-O_2 gas system. *Ceram. Int.* **2016**, *42*, 18373–18379. [CrossRef]
8. Mir, F.A.; Khan, N.Z.; Parvez, S. Recent advances and development in joining ceramics to metals. *Mater. Today Proc.* **2021**, *46*, 6570–6575. [CrossRef]
9. Zhang, Y.; Chen, Y.K.; Yu, D.S.; Sun, D.Q.; Li, H.M. A review paper on the effect of the welding process of ceramics and metals. *J. Mater. Res. Technol.* **2020**, *9*, 16214–16236. [CrossRef]
10. Caron, N.; Bianchi, L.; Méthout, S. Development of a functional sealing layer for SOFC applications. *J. Therm. Spray Technol.* **2008**, *17*, 598–602. [CrossRef]
11. Darsell, J.T.; Weil, K.S. High-temperature strength of YSZ joints brazed with palladium silver copper oxide filler metals. *Int. J. Hydrogen Energy.* **2011**, *36*, 4519–4524. [CrossRef]
12. Hardy, J.S.; Kim, J.Y.; Thomsen, E.C.; Weil, K.S. Improved wetting of mixed ionic/electronic conductors used in electrochemical devices with ternary air braze filler metals. *J. Electrochem. Soc.* **2007**, *154*, 32. [CrossRef]
13. Kiebach, R.; Engelbrecht, K.; Grahl-Madsen, L.; Sieborg, B.; Chen, M.; Hjelm, J.; Norrman, K.; Chatzichristodoulou, C.; Hendriksen, P. An Ag-based brazing system with a tunable thermal expansion for the use as sealant for solid oxide cells. *J. Power Sources* **2016**, *315*, 339–350. [CrossRef]
14. Ayoub, G.; Veljovic, D.; Zebic, M.L.; Miletic, V.; Palcevskis, E.; Petrovic, R.; Janackovic, D. Composite nanostructured hydroxyapatite/yttrium stabilized zirconia dental inserts—The processing and application as dentin substitutes. *Ceram. Int.* **2018**, *44*, 18200–18208. [CrossRef]
15. Tsipas, S.A.; Golosnoy, I.O. Effect of substrate temperature on the microstructure and properties of thick plasma-sprayed YSZ TBCs. *J. Eur. Ceram.* **2011**, *31*, 2923–2929. [CrossRef]
16. Tucker, M.C.; Jacobson, C.P.; De Jonghe, L.; Visco, S. A braze system for sealing metal-supported solid oxide fuel cells. *J. Power Sources* **2006**, *160*, 1049–1057. [CrossRef]
17. Zhou, X.; Sun, K.; Yan, Y.; Le, S.; Zhang, N.; Sun, W.; Wang, P. Investigation on silver electric adhesive doped with Al_2O_3 ceramic particles for sealing planar solid oxide fuel cell. *J. Power Sources* **2009**, *192*, 408–413. [CrossRef]
18. Si, X.; Cao, J.; Talic, B.; Ritucci, I.; Li, C.; Qi, J.; Feng, J.; Kiebach, R. A novel Ag-based sealant for solid oxide cells with a fully tunable thermal expansion. *J. Alloys Compd.* **2020**, *831*, 154608. [CrossRef]
19. Huang, L.W.; Wu, Y.Y.; Shiue, R.K. The effect of oxygen pressure in active brazing 8YSZ and Crofer 22H alloy. *J. Mater. Res. Technol.* **2021**, *10*, 1382–1388. [CrossRef]
20. Huang, L.W.; Shiue, R.K.; Liu, C.K.; Cheng, Y.N.; Lee, R.Y.; Tsay, L.W. Vacuum Brazing of Metallized YSZ and Crofer Alloy Using 72Ag-28Cu Filler Foil. *Materials* **2022**, *15*, 939. [CrossRef]
21. Xin, C.; Liu, W.; Li, N.; Yan, J.; Shi, S. Metallization of Al_2O_3 ceramic by magnetron sputtering Ti/Mo bilayer thin films for robust brazing to Kovar alloy. *Ceram. Int.* **2016**, *42*, 9599–9604. [CrossRef]
22. Song, X.G.; Niu, C.N.; Hu, S.P.; Liu, D.; Cao, J.; Feng, J.C. Contact reactive brazing of Al7075 alloy using Cu layer deposited by magnetron sputtering. *J. Mater. Process. Technol.* **2018**, *252*, 469–476. [CrossRef]
23. Singh, M.; Shpargel, T.P.; Asthana, R. Brazing of yttria-stabilized zirconia (YSZ) to stainless steel using Cu, Ag, and Ti-based brazes. *J. Mater. Sci.* **2008**, *43*, 23–32. [CrossRef]
24. Yang, Z.W.; Xiong, Z.; Wang, J.L.; Wang, Y.; Wang, D.P. Microstructural evolution and high-temperature oxidation resistance of YSZ/Crofer 22H brazed joints using Ag-based filler for solid-oxide fuel cell applications. *Mater. Charact.* **2023**, *200*, 112888. [CrossRef]
25. Lin, P.; Lin, T.; He, P.; Wang, M.; Yang, J. Microstructure evolution and mechanical properties of a vacuum-brazed Al_2O_3/Ti joint with Mo-coating on Al_2O_3 and Ti surfaces. *Ceram. Int.* **2019**, *45*, 11195–11203. [CrossRef]
26. Laik, A.; Mishra, P.; Bhanumurthy, K.; Kale, G.B.; Kashyap, B.P. Microstructural evolution during reactive brazing of alumina to Inconel 600 using Ag-based alloy. *Acta Mater.* **2013**, *61*, 126–138. [CrossRef]
27. Wang, J.L.; Yang, Z.W.; Wang, Y.; Wang, D.P.; Li, H.J. Microstructural stability and mechanical properties of Al_2O_3/Kovar 4 J34 joint vacuum brazed using Ag-5Cu-1Al-1.25Ti (wt%) filler metal. *J. Manuf. Process.* **2021**, *72*, 553–564. [CrossRef]
28. Massalski, T.B. *Binary Alloy Phase Diagrams*, 2nd ed.; ASM International: Materials Park, OH, USA, 1992.
29. Humpston, G.; Jacobson, D.M. *Principles of Soldering and Brazing*; ASM International: Materials Park, OH, USA, 1993.
30. Shiue, R.K.; Wu, S.K.O.J.; Wang, J.Y. Microstructural evolution at the bonding interface during the early-stage infrared active brazing of alumina. *Metall. Mater. Trans.* **2000**, *31*, 2527–2536. [CrossRef]
31. Villars, P.; Prince, A.; Okamoto, H. *Handbook of Ternary Alloy Phase Diagrams*; ASM International: Materials Park, OH, USA, 1995.

materials

MDPI

Article

Electrical Conductivity and Antibacterial Activity of Woven Fabrics through Quercetin-Assisted Thermal Reduction of a Graphene Oxide Coating

Mariia Svyntkivska [1,*], Tomasz Makowski [1], Dorota Kregiel [2] and Ewa Piorkowska [1,*]

[1] Centre of Molecular and Macromolecular Studies, Polish Academy of Sciences, Sienkiewicza 112, 90-363 Lodz, Poland; tomasz.makowski@cbmm.lodz.pl

[2] Department of Environmental Biotechnology, Faculty of Biotechnology and Food Sciences, Lodz University of Technology, Wolczanska 171/173, 90-924 Lodz, Poland; dorota.kregiel@p.lodz.pl

* Correspondence: mariia.svyntkivska@cbmm.lodz.pl (M.S.); ewa.piorkowska@cbmm.lodz.pl (E.P.); Tel.: +48-42-6803-316 (M.S. & E.P.)

Abstract: Cotton and poly(ethylene terephthalate) (PET) woven fabrics were coated with graphene oxide (GO) using a padding method and the GO deposited on the fiber surfaces was thermally reduced to impart electrical conductivity to the fabrics. To assist the thermal reduction of GO, quercetin (Q)—a natural flavonoid—was used. To this end, before the reduction, the GO-padded fabrics were immersed in Q solutions in ethanol with different Q concentrations. Q enhanced the thermal reduction of GO. Depending on the Q concentration in the solutions, electrical surface resistivities of the cotton fabric of 750 kΩ/sq to 3.3 MΩ/sq and of the PET fabric of 240 kΩ/sq to 730 kΩ/sq were achieved. The cotton and PET fabrics also became hydrophobic, with water contact angles of 163° and 147°, respectively. In addition to the electrical conductivity, the presence of Q resulted in antibacterial activity of the fabrics against *Escherichia coli* and *Staphylococcus aureus*.

Keywords: electroconductive fabric; reduced graphene oxide; antibacterial activity; quercetin

1. Introduction

Electrically conductive textiles are materials with a broad range of potential applications, from protective or smart cloth and wearable heaters to sensors, solar cells, and supercapacitors [1–14].

To impart electrical conductivity to synthetic polymers, they can be blended with conducting particles or nanoparticles. In turn, the deposition of conductive particles on fiber surfaces may impart electrical conductivity to natural fibers or ready-made synthetic fibers. Among others, carbon nanoparticles, including graphene and its derivatives (GM), are used for such a purpose [3,14–16]. The application of graphene for the modification of textiles is limited by the difficulty in obtaining stable aqueous dispersions of it, which can be applied for coating hydrophilic fibers, for instance, cotton. In turn, graphene oxide (GO) is hydrophilic and easily dispersible in water because of the presence of functional groups such as hydroxyl, carboxyl, carbonyl and epoxy groups. Epoxide and hydroxyl groups are located on the GO basal plane, whereas carboxyl groups are present on the edges of GO sheets [17]. GO has broad potential applications in biomedicine and nanomedicine because of its unique features, including a 2D planar structure, a large surface area, a high mechanical strength, ease of modification, chemical stability, and biocompatibility. However, GO's functional groups have a negative effect on its electrical conductivity. To improve its conductivity, GO has to be reduced through thermal or chemical treatment to reduced graphene oxide (rGO) This reduction also increases the hydrophobicity and π-π stacking important for use in drug delivery [17,18].

The chemical reduction of GO to rGO requires an elevated temperature, 80–100 °C, and needs a relatively long time, from several hours to days. Thermal reduction demands an even higher temperature and an oxygen-free atmosphere is beneficial [19–21].

In our previous studies, the electrical conductivity of cotton fabric was achieved by coating it with GO and a subsequent thermal GO reduction by heating to 220 °C and short annealing at this temperature. To enhance the reduction, L-ascorbic acid and antioxidants applied in the plastic industry were used [22,23]. Depending on the antioxidant used, the fabric samples became hydrophilic or hydrophobic. It is also of importance that cotton decomposition occurs above 240–250 °C [24].

The growing interest in smart garments, which are electroconductive, sensitive to motion, and able to store energy or communicate, poses additional demands for the textiles used in such applications. In particular, antimicrobial activity is beneficial as it can limit or eradicate microbial colonization and the infection risk. The use of particles of GM offers such a possibility. However, the antibacterial activity of GM depends on many factors, such as the dispersibility, adsorption ability, number of corners and sharp edges [25], and is affected not only by particle shape, lateral size, and layer number but also by modifications of the surface, dispersion, and agglomeration.

Three main mechanisms of antibacterial activity of GM are described in the literature. The main mechanisms of antibacterial activity of rGO are related to the nanoblade effect and oxidative stress [25]. The first one, frequently observed, is related to sharp particle edges, and termed as the nanoblade effect. The nanoblades, called also nanoknives or cutters, cut cell membranes and damage them, resulting in bacteria death [26–28]. The second mechanism is oxidative stress, caused by reactive oxygen species, leading to bacteria DNA damage and mitochondrial dysfunction [29]. The third mechanism, observed in solutions rather than in coatings, is the wrapping or trapping of bacteria by GM platelets due to their flexible structure, which isolates the bacteria from the environment [30]. In our previous studies, the antibacterial activity of GO and rGO coatings of cotton fabric was greatly limited [31]. The flakes, adhering to cotton fibers and lying flat on them, did not expose any sharp edges. Moreover, no wrapping or trapping of bacteria by the flakes was found. It is also worth noting that to impart electroconductivity to the fabric, the amount of rGO particles deposited on the fibers should be sufficient to create a 3D network, but the full coverage of fibers is not required, and the uncoated areas of cotton fibers were prone to bacteria colonization.

In our previous work, to make the rGO-coated fabric antibacterial, Ag particles were deposited electrochemically [31]. Others have obtained electrically conductive and antibacterial rGO/Ag-modified linen and cotton fabrics by means of the simultaneous reduction of GO and silver nitrate with trisodium citrate [32]. To make poly(ethylene terephthalate) (PET) fabric electrically conductive and antibacterial, the simultaneous reduction of GO and silver nitrate was carried out with the use of ascorbic acid [33].

Recently, there has been a growing interest in the use of natural compounds in diverse fields, including the synthesis of nanoparticles [34] and reduction of GO [35,36], especially considering that among chemical reducing agents, the most efficient reductant is hydrazine, which is toxic and harmful to the environment. The use of natural reductants is especially important for the reduction of GO deposited on natural fibers, where the use of harmful compounds is excluded. Flavonoids, belonging to natural polyphenols, are antioxidants [37] and also exhibit antibacterial activity [38]. Quercetin (Q) is a natural flavonoid that occurs in plants, plant-derived foods, and nutritional supplement products. It exhibits anticancer, anti-inflammatory, antiallergy, antibacterial, and antiviral properties [39–42]. A novel drug delivery system based on GO loaded with Q was recently developed [43]. Q was also utilized to chemically reduce GO in order to cast Q-rGO on paraffin impregnated graphite electrodes for analysis of metal ions [44].

According to our knowledge, Q has not been used to reduce GO coatings on a textile. Being a natural flavonoid, Q is a promising eco-friendly reducing agent for GO deposited on fibers, including natural fibers. In the present study, Q was applied to assist the thermal reduction of GO deposited on cotton and PET fabrics. The fabrics were immersed in Q solutions in ethanol and heated to reduce the GO. This treatment allowed to achieve electrical conductivity of the fabrics, dependent on Q concentration in the solutions. Electrical surface resistivities of 750 kΩ/sq and 240 kΩ/sq for the cotton and PET fabrics, respectively, were reached. After the reduction, the fabrics were hydrophobic. Moreover, the presence of Q imparted antibacterial activity against *Escherichia coli* and *Staphylococcus aureus* to the fabrics.

2. Materials and Methods

2.1. Materials

White plain weave cotton fabric (145 g/m^2) of 0.36 mm thickness with a weft of 205 threads/10 cm and a warp of 295 threads/10 cm was used in this study. Before further experiments, the fabric was cleaned by diethyl ether and anhydrous ethanol extraction, as previously described [45]. Also, plain weave PET fabric (Wistil, Kalisz, Poland) (89 g/m^2) of 0.19 mm thickness was used, with a warp of 390 threads/10 cm, and a weft of 320 threads/10 cm.

A 1 wt.% aqueous dispersion of GO was supplied by NanoCarbon (Warsaw, Poland). According to the supplier, the I_D/I_G value was >1.87 based on Raman spectroscopy. Elemental analysis evidenced the following element contents (%*w*/*w*): C, >45%; H, <2.5%; N, <0.5%; O, <49%; other, <4%. An impurity analysis by X-ray fluorescence spectrometry yielded: S (2.5%) > Ca (1%) > Mn (0.5%) > K (0.3%) > Cl (0.08%) = Fe (0.08%) > Cu (0.02%) = Zn (0.02%) > Ni (0.007%) > Cr (0.006%). Before further use, to reduce the GO concentration to 0.1 wt.%, the dispersion was diluted with distilled water. Then, it was subjected to sonication using a Sonopuls HD 2200 ultrasonic homogenizer (Bandelin, Berlin, Germany) at room temperature (RT) for 30 min (amplitude 40%, frequency 20 kHz).

Anhydrous Q (3,3′,4′,5,7-pentahydroxyflavone, $C_{15}H_{10}O_7$), a flavonoid with a molar mass of 302.24 g/mol, was purchased from Apollo Scientific (Stockport, UK). Ethanol and diethyl ether were purchased from POCH (Gliwice, Poland) and CHEMPUR (Piekary Slaskie, Poland), respectively.

2.2. Modification of Fabrics

The fabrics were padded with the GO dispersion using a laboratory double-roll padding machine with horizontal squeezing rollers at a rate of 0.01 m/min, and then dried in the air at 100 °C for 15 min. The padding and drying cycle was repeated four times, as described previously in [22,23,45].

Strips of fabrics coated with GO were immersed in solutions of anhydrous Q in ethanol with concentrations of 0.25, 0.5, and 1 wt.% for 5 min, and then dried at RT. It should be noted that the solubility of Q in ethanol in room conditions is approx. 1 wt.% [46].

To reduce GO, the fabric strips, both strips treated with Q and untreated strips, placed between microscope glasses were thermally treated in a hot-stage FP82 equipped with an FP90 temperature controller from Mettler Toledo (Greifensee, Switzerland). They were heated from RT to 220 °C, annealed at 220 °C for 1 min, and then cooled down to RT at 10 °C/min.

The obtained materials are referred to as PrGO, CrGO, PrGO/Q, and CrGO/Q, where P and C denote PET and cotton, respectively, and Q denotes Q-assisted reduction of GO. The number following the letter Q, for example, PrGO/Q0.5 or CrGO/Q0.5, indicates the percentage of Q in ethanol.

2.3. Characterization

Fabric samples, before and after modification, were vacuum-sputtered with 10 nm gold layers using a Quorum EMS150R ES (Laughton, UK) and examined with a scanning electron microscope (SEM), JEOL 6010LA (Tokyo, Japan).

The thermal properties of the PET fabric before and after thermal treatment were studied using a DSC3 differential scanning calorimeter from Mettler Toledo (Greifensee, Switzerland) during heating at a rate of 10 °C/min to 300 °C in a nitrogen atmosphere.

The decrease in surface electrical resistivity (R) of the samples during the reduction was monitored via the 2-wire method, whereas after 24 h of stabilization under room conditions, R was determined via the 4-wire method using a Keithley 2400C SourceMeter (Cleveland, OH, USA). Copper electrodes were attached to the specimen surfaces at 1 cm apart with Dotite D-550 silver-containing paste (Fujikura Kasei, Tokyo, Japan). The results were averaged over at least three measurements of each material. Electrical impedance measurements were carried out at RT in a frequency range of 20 Hz–1 MHz by a Hewlett Packard 4284A Precision LCR Meter (Palo Alto, CA, USA). Two copper electrodes were attached to the specimen surface with silver paint at a distance of 1 cm apart. A 1 V amplitude of the sinusoidal voltage was maintained.

The water contact angles (WCAs) of the fabric samples were measured using the drop method with 5 μL drops at RT. A 100–00-230 NRL Rame Hart goniometer (Succasunna, NJ, USA) and the ImageJ Drop Analysis program (1.46r) were used for this purpose. The measurements were carried out five times and average WCA values were calculated.

The antibacterial properties of the fabric samples against the Gram-positive bacterial strain *Staphylococcus aureus* (*S. aureus*) ATCC 6538 and the Gram-negative bacterial strain *Escherichia coli* (*E. coli*) ATCC 8739 were analyzed according to Antibacterial Activity Assessment of Textile Materials: Agar Diffusion Test (ISO 20645:2004 [47]). At first, 15 mL portions of nutritive agar medium (Trypticase Soy Agar, Merck, Darmstadt, Germany) were poured onto sterile Petri dishes. The inoculum of a bacterial tested culture of 1–5 $\times$ 10^8 CFU/mL (0.2 mL) was then poured onto the agar plates. Then, fabric specimens in the form of disks with a 6 mm diameter were placed on the agar surface. After 24 h incubation at 37 °C, the contact zones under the specimens were visually analyzed, and then the bacterial growth inhibition zones, if present, were measured. To evaluate the bacterial colonization, the disk specimens removed from the agar were also analyzed by SEM.

3. Results and Discussion

3.1. Graphene Oxide Reduction and Electrical Resistivity of Modified Fabrics

Figure 1 shows cotton and PET fabric samples before and after coating with GO and after thermal reduction of GO. Coating with GO made the fabrics brown. After the thermal reduction of GO, the fabrics darkened.

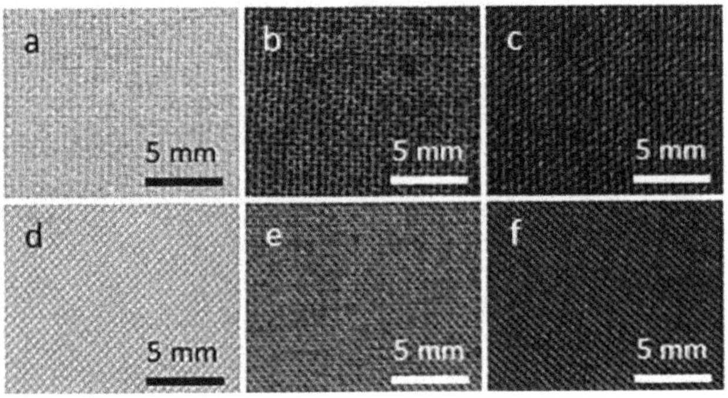

Figure 1. Cotton (**a**–**c**) and PET (**d**–**f**) fabric samples before coating with GO (**a**,**d**), after coating with GO (**b**,**e**), and after thermal reduction of GO (**c**,**f**).

Figures 2 and 3 show SEM micrographs of the cotton fabric and the PEF fabric, respectively, before and after deposition of GO, after immersion in 1 wt.% solution of Q in ethanol, and after thermal reduction of GO.

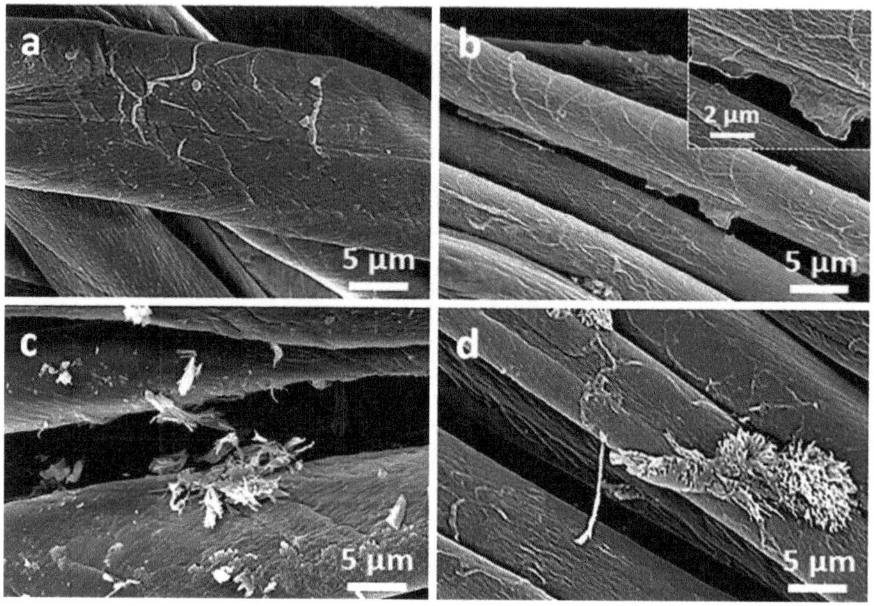

Figure 2. SEM micrographs of cotton fabric before (**a**) and after deposition of GO (**b**), after immersion in 1 wt.% solution of Q in ethanol (**c**), and after thermal reduction of GO (CrGO/Q1) (**d**).

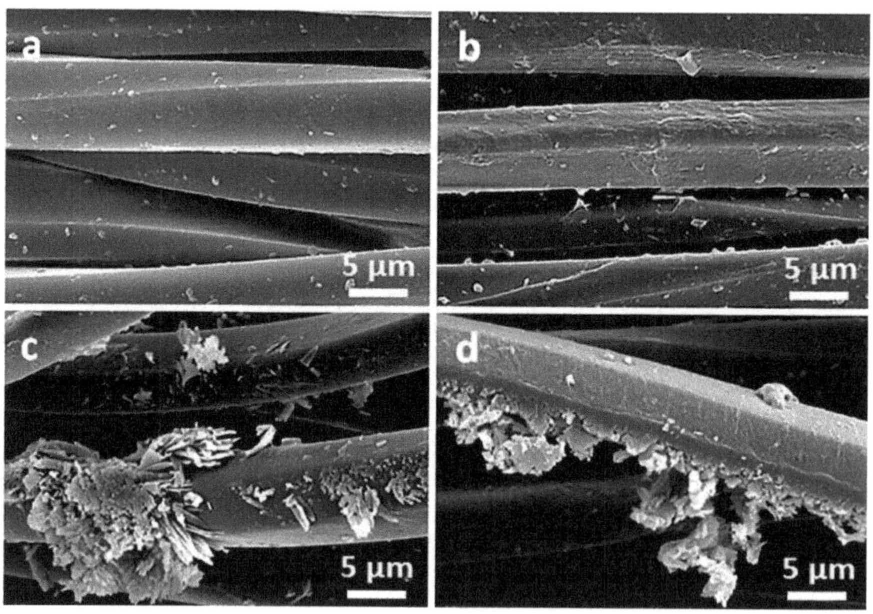

Figure 3. SEM micrographs of PET fabric before (**a**) and after deposition of GO (**b**), after immersion in 1 wt.% solution of Q in ethanol (**c**), and after thermal reduction of GO (PrGO/Q1) (**d**).

Due to their small thickness and good dispersion, GO or rGO flakes were only occasionally discernible on the rough cotton fibers, as shown in the inset in Figure 2b. The flakes were better seen on smooth PET fibers, as illustrated in Figure 3b.

On the contrary, on the fibers immersed in a Q solution in ethanol, Q crystals were easily observed before and after the thermal reduction of GO, as illustrated in Figures 2c,d and 3c,d. It is worth noting that the thermal treatment did not alter the cotton and PET fiber surface morphology.

Moreover, thermal treatment did not induce any significant changes in the thermal properties of PET fabric, as illustrated in Figure 4. The heating thermogram of the untreated PET fabric exhibited a melting peak at 255 °C with an enthalpy of 57 J/g. In addition, it exhibited a very weak and broad melting endotherm followed by a cold crystallization exotherm with extrema at 163 °C and 208 °C, respectively. The small enthalpies associated with these transitions, about 3–4 J/g, were similar. After the thermal treatment analogous to that used for GO reduction, and also after Q-assisted rGO reduction (PrGO/Q1), a low-temperature endotherm peak appeared at 224 °C, with an enthalpy of 1.5 J/g, whereas cold crystallization was absent. However, the main melting peak temperature remained unchanged, and its enthalpy only insignificantly decreased by 2–3 J/g. Thus, the crystallinity of PET, calculated assuming the heat of fusion of crystals is 133–140 J/g [48,49], remained practically the same at near 40%.

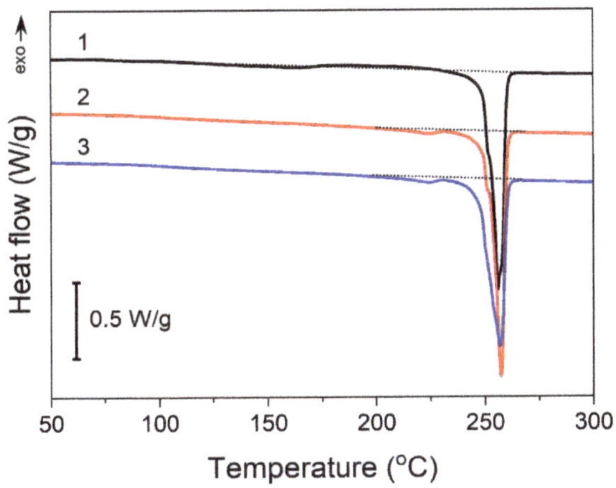

Figure 4. DSC heating thermograms of PET fabric before thermal treatment (1), after thermal treatment analogous to that during thermal reduction of GO (2), and (3) after thermal reduction of GO (PrGO/Q1). Heating rate of 10 °C/min.

During heating and annealing at 220 °C, the R of GO-coated fabrics decreased due to the reduction of GO as shown in Figure 5. However, R increased during cooling and further holding at RT, but stabilized after 24 h. The resistivity of GM particles is known to increase with decreasing temperatures [50]; however, such a dependence could not be responsible for the several-fold increase in R of the fabrics, especially considering that the increase also occurred at RT. Such an increase in R was observed and discussed by us previously [22], and was attributed to the partial reversibility of the reduction. The values of R of the fabrics measured 24 h after the reduction are shown in Figure 6a and Table 1. The R values of the fabrics untreated with Q were the highest, and that of PrGO was above the measurement range of the equipment, 50 MΩ. The current–voltage characteristics of all samples, shown in Figure 6b, were linear, evidencing that Ohm's law was satisfied.

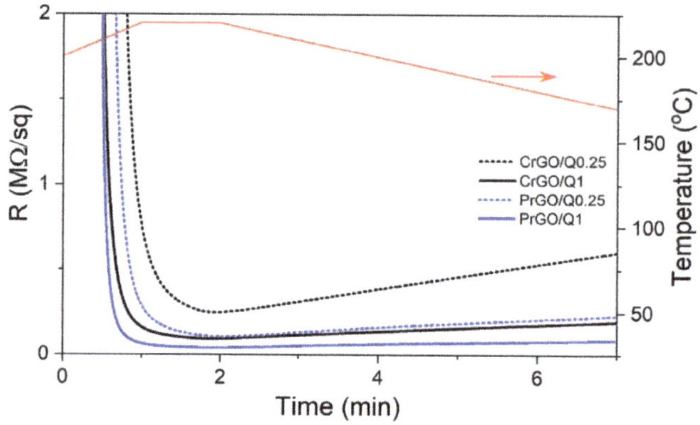

Figure 5. Exemplary dependencies of electrical surface resistivity, R, of the GO-coated cotton and PET fabric on time during Q-assisted thermal reduction of GO (black and blue lines) and temperature dependence on time (red line).

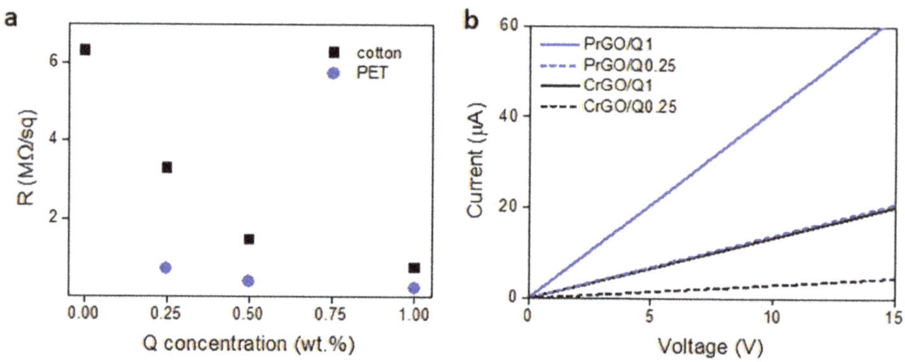

Figure 6. Electrical surface resistivity, R, of cotton and PET fabrics after thermal reduction of GO depending on Q concentration in its solution in ethanol (**a**), and exemplary current–voltage characteristics measured 24 h after thermal reduction of GO (**b**).

Table 1. Electrical surface resistivity, R, and water contact angles, WCAs, of cotton and PET fabrics after thermal reduction of GO, depending on Q concentration in solution in ethanol.

Concentration of Q (%)	Cotton		PET	
	R (MΩ/sq)	WCA (°)	R (MΩ/sq)	WCA (°)
-	6.3	155	>50	144
0.25	3.3	163	0.73	146
0.5	1.4	163	0.40	147
1	0.75	163	0.24	147

Cellulose is a polymer composed of D-glucose units linked through covalent β-1,4-glycosidic bonds. We hypothesize that the glucose units could assist the thermal reduction of GO, as glucose is a mild reducing agent for GO and was recently used to synthesize glucose-rGO-supported Ag-Cu_2O nanocomposites [51]. The application of Q to assist the reduction of GO permitted a reduction in the R of the fabrics; it decreased with increasing concentrations of Q in ethanol solutions, as shown in Figure 6a, although the dependence was weaker for larger Q concentrations. At 1 wt.% Q concentration, an R of 750 kΩ/sq and 240 kΩ/sq was achieved for cotton (CrGO/Q1) and PET (PrGO/Q1) fabrics, respectively.

However, the R values of CrGO/Q were higher than those of PrGO/Q. We hypothesize that after the strongly enhanced reduction of GO, the surface morphology of the fibers becomes an important factor. The surface roughness of cotton fibers, illustrated in Figure 2, could negatively affect the contacts between rGO platelets and this could result in a higher resistance of the rGO-conducting network on fiber surfaces. It is worth mentioning that an adverse effect of polylactide surface roughness on the formation of a conducting network of multi-wall carbon nanotubes (MWCNTs) was observed by us previously [52].

Frequency dependencies of the electrical impedance modulus of CrGO/Q1 and PrGO/Q1 are plotted in Figure 7a. The corresponding Nyquist plots of the real and imaginary parts of the complex impedances are shown in Figure 7b. They are in the form of semicircular arcs, with diameters corresponding to the fabric resistances. In a broad frequency range, the fabric impedances were dominated by their frequency-independent resistances, whose values corresponded to R values determined independently in DC experiments. At higher frequencies, transitions to a frequency-dependent regime dominated by a negative reactance, that is, capacitance, were observed. For the material with a higher low-frequency impedance and R, the frequency at which the transition occurred was lower. Similar behavior was observed by us previously [52] for polylactide nonwovens modified with MWCNTs. The frequency-independent impedance was interpreted as being determined by direct contacts between the MWCNTs forming the conducting network. The impedance spectroscopy results confirmed the dominant role of direct rGO contacts on the electrical properties of rGO-modified fabrics in a broad frequency range.

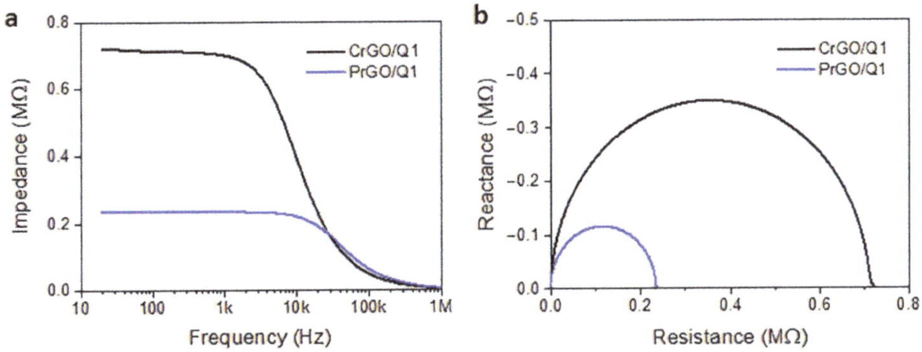

Figure 7. Electrical impedance modulus of CrGO/Q1 and PrGO/Q1 (**a**) and Nyquist plots of reactance vs. resistance (**b**).

Figure 8 shows water droplets on the fabric surfaces, whereas the corresponding WCA values are collected in Table 1. Both the cotton and PEF fabrics were hydrophilic and water droplets soaked into them, although this was slower in the PET fabric than in the cotton fabric. The GO-coated fabrics behaved similarly. However, after the thermal reduction of GO, all fabric samples became hydrophobic. WCAs of 155° and 144° were measured for CrGO and PrGO, respectively, which increased to 163° and 147° for CrGO/Q and PrGO/Q. It seems that the surface roughness of cotton fibers, which adversely affected the R of CrGO/Q, improved their hydrophobicity, as the air trapped in pockets and between protrusions of rGO-coated cotton fibers could enhance the hydrophobicity of the fabric [53].

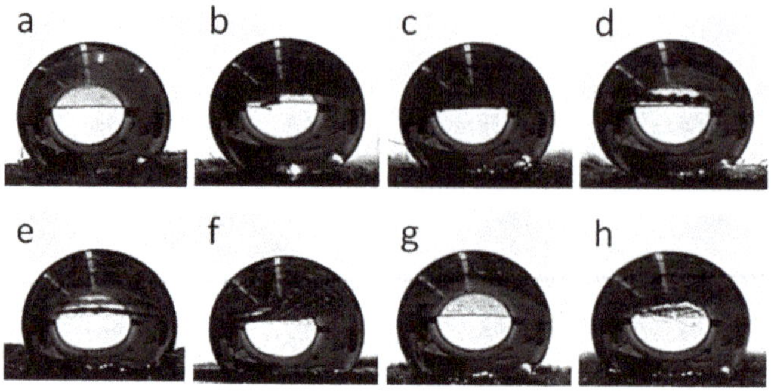

Figure 8. Water droplets deposited on fabric surfaces: CrGO (**a**), CrGO/Q0.25 (**b**), CrGO/Q0.5 (**c**), CrGO/Q1 (**d**), PrGO (**e**), PrGO/Q0.25 (**f**), PrGO/Q0.5 (**g**), and PrGO/Q1 (**h**).

3.2. Antibacterial Properties

Exemplary photographs of the fabric specimens lying on inoculated agar after bacteria incubation are shown in Figure 9. Bacterial growth inhibition zones were barely noticeable, but a clearly reduced growth close to the tested materials was visible. The zone was only larger around CrGO/Q1 on *S. aureus*-inoculated agar, with a diameter of 6.6 mm. However, the absence of bacterial growth inhibition zones around specimens does not exclude their antibacterial activity in direct contact with microorganisms. The antibacterial activity of the CrGO/Q and PrGO/Q coated fabric specimens was corroborated by SEM analysis of their surfaces, which were in contact with the inoculated agar. Both cotton and PET fabrics, shown in Figures 10 and 11, were overgrown with *E. coli* and *S. aureus*. The same applies to the fabrics coated with GO. It seems that slightly less bacteria were seen on CrGO and PrGO, most possibly due to the hydrophobicity of these materials, which made the adhesion of bacteria more difficult. However, the bacterial colonization was significantly reduced only on CrGO/Q and PrGO/Q. The amount of bacteria seen on CrGO/Q0.25 and PrGO/Q0.25 was significantly decreased in comparison to that on the materials untreated with Q, as shown in Figures 10c,g and 11c,g. Similar results were obtained for CrGO/Q0.5 and PrGO/Q0.5. However, the strongest antibacterial activity was evidenced for CrGO/Q1 and PrGO/Q1, on which the bacterial growth was strongly reduced. Moreover, many of these bacterial cells were flattened and damaged. Such changes have been attributed by others to bacteria damage and death caused by contact with antimicrobial materials [54–56].

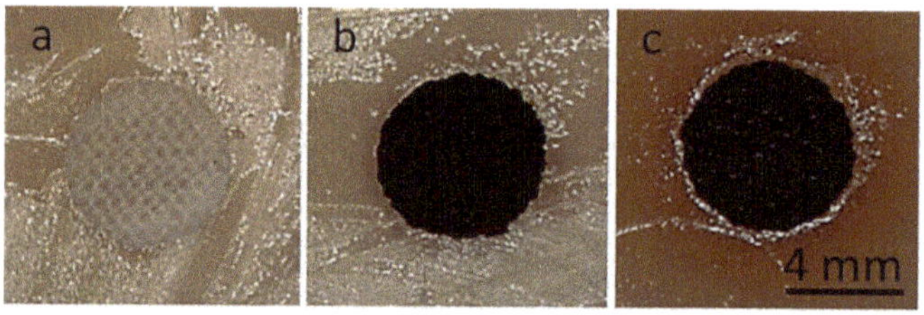

Figure 9. Cotton fabric disks with a 6 mm diameter in *S. aureus*-inoculated agar after 24 h of incubation: neat cotton (**a**), CrGO (**b**), CrGO/Q1 (**c**).

It is worth noting that Q decomposes during heating. According to [57], the first stage of decomposition in air occurs between 103 and 342 °C, although mass loss in this temperature range is only 10%. The obtained results evidenced that despite its minor decomposition, the antibacterial activity of Q did not vanish after the thermal treatment and reduction of GO.

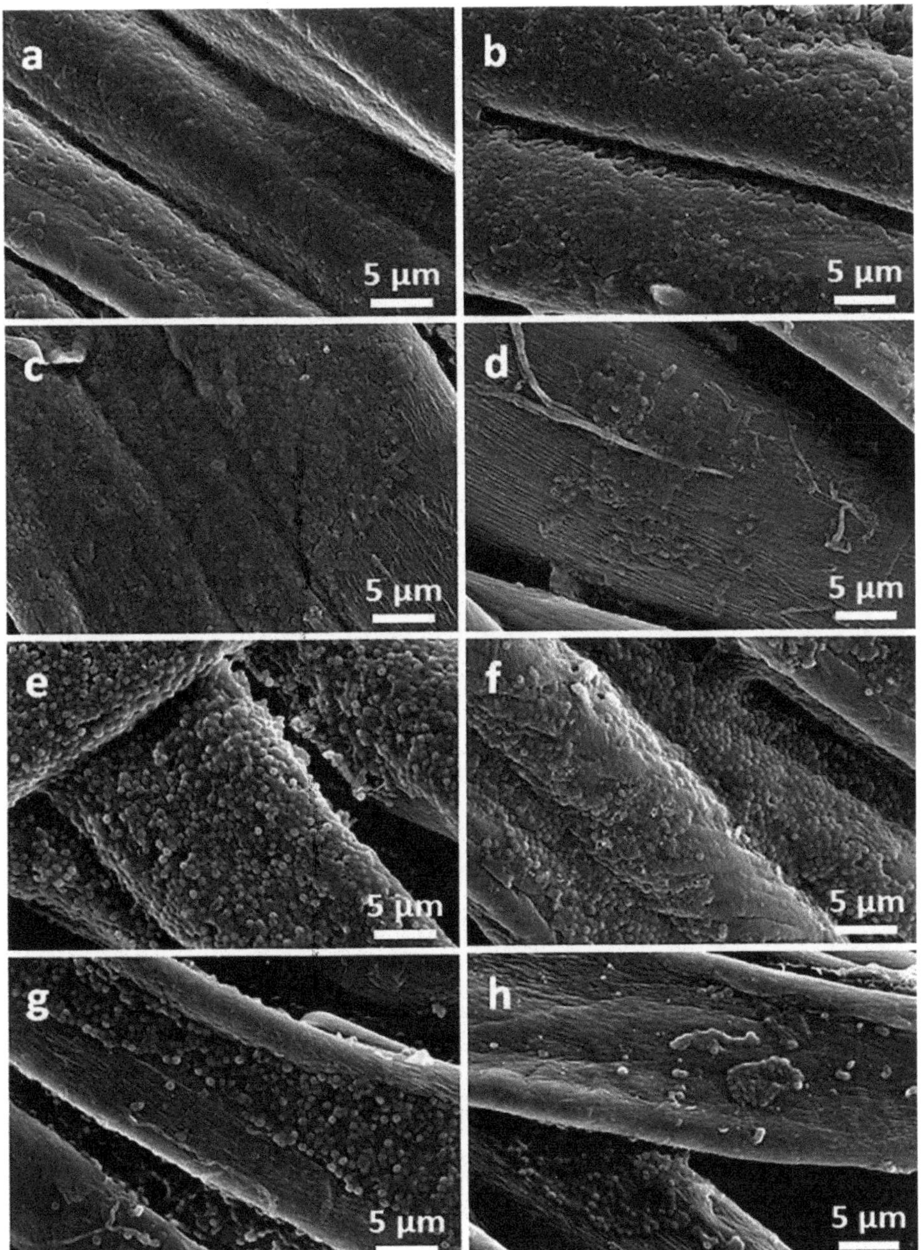

Figure 10. SEM micrographs of bottom sides of cotton fabric specimens removed from *E. coli* (**a**–**d**) and *S. aureus* (**e**–**h**) inoculated agar: pure cotton (**a**,**e**), CrGO (**b**,**f**), CrGO/Q0.25 (**c**,**g**), and CrGO/Q1 (**d**,**h**).

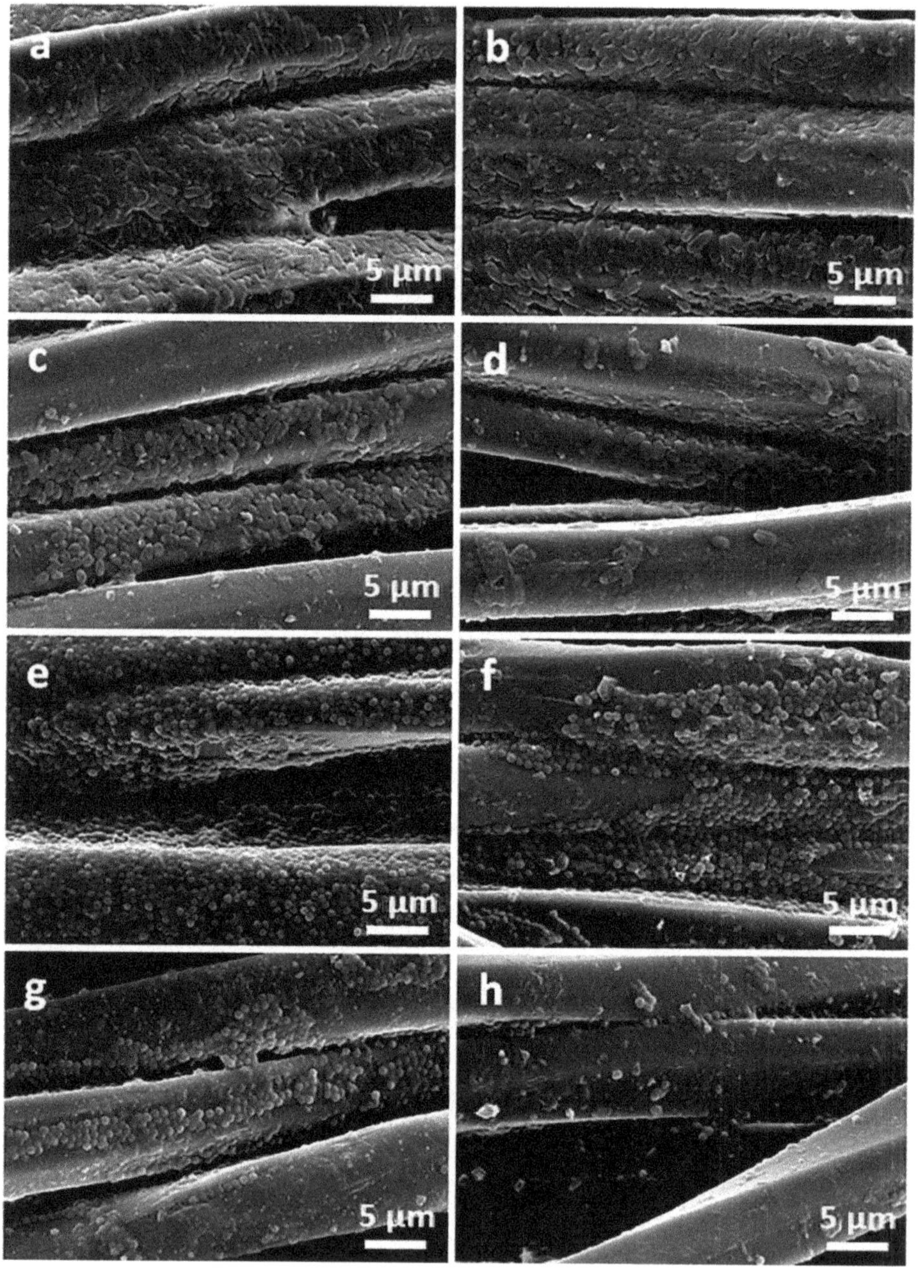

Figure 11. SEM micrographs of bottom sides of PET fabric specimens removed from *E. coli* (**a**–**d**) and *S. aureus* (**e**–**h**) inoculated agar: pure PET (**a**,**e**), PrGO (**b**,**f**), PrGO/Q0.25 (**c**,**g**), and PrGO/Q1 (**d**,**h**).

4. Conclusions

Cotton and PET fabrics were coated with a GO dispersion using the padding method, and then GO was reduced thermally to rGO with the assistance of quercetin, Q, a natural flavonoid. The use of Q enhanced the reduction of GO, leading to a diminished electrical surface resistivity of the materials, R. The R decreased with increasing concentration of Q in an ethanol solution, in which the GO-coated fabrics were immersed before the thermal reduction of GO. At 1 wt.% Q concentration, an R of 750 kΩ/sq and 240 kΩ/sq was achieved for cotton and PET fabric, respectively. The higher R of cotton fabrics was most probably related to the roughness of cotton fibers having an adverse effect on the formation of the conducting network of rGO platelets on fiber surfaces. After the reduction of the GO deposited on the fibers, both cotton and PET fabrics became hydrophobic. Q-assisted reduction of GO resulted in WCA values of 163° and 147°, respectively. Moreover, treatment with Q resulted in antibacterial activity of the materials against *E. coli* and *S. aureus*. SEM analysis proved the reduced bacteria colonization on all Q-treated fibers. The strongest effect was evidenced for the fabrics immersed in a 1 wt.% solution of Q in ethanol before GO reduction, which agreed with the presence of bacterial growth inhibition zone for CrGO/Q1. The combination of electrical conductivity, hydrophobicity, and antibacterial activity makes the cotton and PET fabrics coated with rGO obtained by Q-assisted reduction of GO interesting materials for a broad range of applications.

Author Contributions: M.S.: conceptualization, investigation, data curation, writing—original draft preparation, funding acquisition; T.M.: conceptualization, formal analysis, writing—review and editing; D.K.: investigation; E.P.: conceptualization, writing—original draft preparation, writing—review and editing, supervision. All authors have read and agreed to the published version of the manuscript.

Funding: This research was funded by National Science Centre (Narodowe Centrum Nauki), Poland, grant number 2020/37/N/ST5/04041 and statutory funds of CMMS PAS.

Institutional Review Board Statement: Not applicable.

Informed Consent Statement: Not applicable.

Data Availability Statement: The data are available on request from the corresponding authors.

Conflicts of Interest: The authors declare no conflict of interest.

References

1. Islam, G.M.N.; Ali, A.; Collie, S. Textile sensors for wearable applications: A comprehensive review. *Cellulose* **2020**, *27*, 6103–6131. [CrossRef]
2. Maity, S.; Chatterjee, A. Conductive polymer-based electro-conductive textile composites for electromagnetic interference shielding: A review. *J. Ind. Text.* **2018**, *47*, 2228–2252. [CrossRef]
3. Molina, J. Graphene-based fabrics and their applications: A review. *RSC Adv.* **2016**, *6*, 68261–68291. [CrossRef]
4. Stoppa, M.; Chiolerio, A. Wearable Electronics and Smart Textiles: A Critical Review. *Sensors* **2014**, *14*, 11957–11992. [CrossRef]
5. Kowalczyk, D.; Fortuniak, W.; Mizerska, U.; Kaminska, I.; Makowski, T.; Brzezinski, S.; Piorkowska, E. Modification of cotton fabric with graphene and reduced graphene oxide using sol-gel method. *Cellulose* **2017**, *24*, 4057–4068. [CrossRef]
6. Wang, D.; Li, D.; Zhao, M.; Xu, Y.; Wei, Q. Multifunctional wearable smart device based on conductive reduced graphene oxide/polyester fabric. *Appl. Surf. Sci.* **2018**, *454*, 218–226. [CrossRef]
7. Seyedin, S.; Razal, J.M.; Innis, P.C.; Jeiranikhameneh, A.; Beirne, S.; Wallace, G.G. Knitted Strain Sensor Textiles of Highly Conductive All-Polymeric Fibers. *ACS Appl. Mater. Interfaces* **2015**, *7*, 21150–21158. [CrossRef] [PubMed]
8. Zahid, M.; Papadopoulou, E.L.; Athanassiou, A.; Bayer, I.S. Strain-responsive mercerized conductive cotton fabrics based on PEDOT:PSS/graphene. *Mater. Design* **2017**, *135*, 213–222. [CrossRef]
9. Hong, S.; Lee, H.; Lee, J.; Kwon, J.; Han, S.; Suh, Y.D.; Cho, H.; Shin, J.; Yeo, J.; Ko, S.H. Highly Stretchable and Transparent Metal Nanowire Heater for Wearable Electronics Applications. *Adv. Mater.* **2015**, *27*, 4744–4751. [CrossRef]
10. Jost, K.; Perez, C.R.; McDonough, J.K.; Presser, V.; Heon, M.; Dion, G.; Gogotsi, Y. Carbon coated textiles for flexible energy storage. *Energy Environ. Sci.* **2011**, *4*, 5060–5067. [CrossRef]
11. Attia, N.F.; El Ebissy, A.A.; Hassan, M.A. Novel synthesis and characterization of conductive and flame retardant textile fabrics. *Polym. Adv. Technol.* **2015**, *26*, 1551–1557. [CrossRef]
12. Nooralian, Z.; Parvinzadeh Gashti, M.; Ebrahimi, I. Fabrication of a multifunctional graphene/polyvinylphosphonic acid/cotton nanocomposite via facile spray layer-by-layer assembly. *RSC Adv.* **2016**, *6*, 23288–23299. [CrossRef]
13. Huang, Y.; Zhu, M.; Pei, Z.; Xue, Q.; Huang, Y.; Zhi, C. A shape memory supercapacitor and its application in smart energy storage textiles. *J. Mater. Chem. A* **2016**, *4*, 1290–1297. [CrossRef]
14. Sahito, I.A.; Sun, K.C.; Arbab, A.A.; Qadir, M.B.; Choi, Y.S.; Jeong, S.H. Flexible and conductive cotton fabric counter electrode coated with graphene nanosheets for high efficiency dye sensitized solar cell. *J. Power Sources* **2016**, *319*, 90–98. [CrossRef]

15. Makowski, T.; Zhang, C.; Olah, A.; Piorkowska, E.; Baer, E.; Kregiel, D. Modification of dual-component fibrous materials with carbon nanotubes and methyltrichlorosilane. *Mater. Design* **2019**, *162*, 219–228. [CrossRef]
16. Shateri-Khalilabad, M.; Yazdanshenas, M.E. Fabricating electroconductive cotton textiles using graphene. *Carbohydr. Polym.* **2013**, *96*, 190–195. [CrossRef]
17. Bellier, N.; Baipaywad, P.; Ryu, N.; Lee, J.Y.; Park, H. Recent biomedical advancements in graphene oxide- and reduced graphene oxide-based nanocomposite nanocarriers. *Biomater. Res.* **2022**, *26*, 65. [CrossRef]
18. Guex, L.G.; Sacchi, B.; Peuvot, K.F.; Andersson, R.L.; Pourrahimi, A.M.; Ström, V.; Farris, S.; Olsson, R.T. Experimental review: Chemical reduction of graphene oxide (GO) to reduced graphene oxide (rGO) by aqueous chemistry. *Nanoscale* **2017**, *9*, 9562–9571. [CrossRef]
19. Chua, C.K.; Pumera, M. Chemical reduction of graphene oxide: A synthetic chemistry viewpoint. *Chem. Soc. Rev.* **2014**, *43*, 291–312. [CrossRef]
20. Pei, S.; Cheng, H.-M. The reduction of graphene oxide. *Carbon* **2012**, *50*, 3210–3228. [CrossRef]
21. Huh, S.H. Thermal Reduction of Graphene Oxide. In *Physics and Applications of Graphene*; Mikhailov, S., Ed.; IntechOpen: Rijeka, Croatia, 2011; pp. 73–90.
22. Makowski, T.; Svyntkivska, M.; Piorkowska, E.; Mizerska, U.; Fortuniak, W.; Kowalczyk, D.; Brzezinski, S. Conductive and super-hydrophobic cotton fabric through pentaerythritol tetrakis(3-(3,5-di-tert-butyl-4-hydroxyphenyl)propionate) assisted thermal reduction of graphene oxide and modification with methyltrichlorosilane. *Cellulose* **2018**, *25*, 5377–5388. [CrossRef]
23. Jedrzejczyk, M.; Makowski, T.; Svyntkivska, M.; Piorkowska, E.; Mizerska, U.; Fortuniak, W.; Brzezinski, S.; Kowalczyk, D. Conductive cotton fabric through thermal reduction of graphene oxide enhanced by commercial antioxidants used in the plastics industry. *Cellulose* **2019**, *26*, 2191–2199. [CrossRef]
24. Matheus, P.; Vinícios, P.; Ademir, J.Z. Structural Characteristics and Thermal Properties of Native Cellulose. In *Cellulose*; van de Ven, T., Godbout, L., Eds.; IntechOpen: Rijeka, Croatia, 2013; pp. 45–68.
25. Cao, G.; Yan, J.; Ning, X.; Zhang, Q.; Wu, Q.; Bi, L.; Zhang, Y.; Han, Y.; Guo, J. Antibacterial and antibiofilm properties of graphene and its derivatives. *Colloid Surf. B-Biointerfaces* **2021**, *200*, 111588. [CrossRef]
26. Nasirzadeh, N.; Azari, M.R.; Rasoulzadeh, Y.; Mohammadian, Y. An assessment of the cytotoxic effects of graphene nanoparticles on the epithelial cells of the human lung. *Toxicol. Ind. Health* **2019**, *35*, 79–87. [CrossRef] [PubMed]
27. Pulingam, T.; Thong, K.L.; Appaturi, J.N.; Nordin, N.I.; Dinshaw, I.J.; Lai, C.W.; Leo, B.F. Synergistic antibacterial actions of graphene oxide and antibiotics towards bacteria and the toxicological effects of graphene oxide on human epidermal keratinocytes. *Eur. J. Pharm. Sci.* **2020**, *142*, 105087. [CrossRef]
28. Zainal-Abidin, M.H.; Hayyan, M.; Ngoh, G.C.; Wong, W.F. From nanoengineering to nanomedicine: A facile route to enhance biocompatibility of graphene as a potential nano-carrier for targeted drug delivery using natural deep eutectic solvents. *Chem. Eng. Sci.* **2019**, *195*, 95–106. [CrossRef]
29. Khan, B.; Adeleye, A.S.; Burgess, R.M.; Russo, S.M.; Ho, K.T. Effects of graphene oxide nanomaterial exposures on the marine bivalve, Crassostrea virginica. *Aquat. Toxicol.* **2019**, *216*, 105297. [CrossRef]
30. Akhavan, O.; Ghaderi, E.; Esfandiar, A. Wrapping Bacteria by Graphene Nanosheets for Isolation from Environment, Reactivation by Sonication, and Inactivation by Near-Infrared Irradiation. *J. Phys. Chem. B* **2011**, *115*, 6279–6288. [CrossRef]
31. Makowski, T.; Svyntkivska, M.; Piorkowska, E.; Mizerska, U.; Fortuniak, W.; Kowalczyk, D.; Brzezinski, S.; Kregiel, D. Antibacterial Electroconductive Composite Coating of Cotton Fabric. *Materials* **2022**, *15*, 1072. [CrossRef]
32. Farouk, A.; Saeed, S.E.-S.; Sharaf, S.; Abd El-Hady, M.M. Photocatalytic activity and antibacterial properties of linen fabric using reduced graphene oxide/silver nanocomposite. *RSC Adv.* **2020**, *10*, 41600–41611. [CrossRef]
33. Moazami, A.; Montazer, M.; Dolatabadi, M.K. Tunable functional properties on polyester fabric using simultaneous green reduction of graphene oxide and silver nitrate. *Fiber. Polym.* **2016**, *17*, 1359–1370. [CrossRef]
34. Mirza, K.; Naaz, F.; Ahmad, T.; Manzoor, N.; Sardar, M. Development of Cost-Effective, Ecofriendly Selenium Nanoparticle-Functionalized Cotton Fabric for Antimicrobial and Antibiofilm Activity. *Fermentation* **2023**, *9*, 18. [CrossRef]
35. Manchala, S.; Tandava, V.S.R.K.; Jampaiah, D.; Bhargava, S.K.; Shanker, V. Novel and Highly Efficient Strategy for the Green Synthesis of Soluble Graphene by Aqueous Polyphenol Extracts of Eucalyptus Bark and Its Applications in High-Performance Supercapacitors. *ACS Sustain. Chem. Eng.* **2019**, *7*, 11612–11620. [CrossRef]
36. De Silva, K.K.H.; Huang, H.H.; Joshi, R.K.; Yoshimura, M. Chemical reduction of graphene oxide using green reductants. *Carbon* **2017**, *119*, 190–199. [CrossRef]
37. Abbas, M.; Saeed, F.; Anjum, F.M.; Afzaal, M.; Tufail, T.; Bashir, M.S.; Ishtiaq, A.; Hussain, S.; Suleria, H.A.R. Natural polyphenols: An overview. *Int. J. Food Prop.* **2017**, *20*, 1689–1699. [CrossRef]
38. Shamsudin, N.F.; Ahmed, Q.U.; Mahmood, S.; Ali Shah, S.A.; Khatib, A.; Mukhtar, S.; Alsharif, M.A.; Parveen, H.; Zakaria, Z.A. Antibacterial Effects of Flavonoids and Their Structure-Activity Relationship Study: A Comparative Interpretation. *Molecules* **2022**, *27*, 1149. [CrossRef]
39. Kobylińska, A.; Janas, K.M. Prozdrowotna rola kwercetyny obecnej w diecie człowieka—Health—Promoting effect of quercetin in human diet. *Postepy Hig. Med. Dosw.* **2015**, *69*, 51–62. [CrossRef]
40. Kobylińska, A. Exogenous quercetin as a proliferation stimulator in tobacco BY-2 cells. *J. Elem.* **2017**, *22*, 245–258. [CrossRef]
41. Gomathi, K.; Gopinath, D.; Rafiuddin Ahmed, M.; Jayakumar, R. Quercetin incorporated collagen matrices for dermal wound healing processes in rat. *Biomaterials* **2003**, *24*, 2767–2772. [CrossRef]

42. Kost, B.; Svyntkivska, M.; Brzeziński, M.; Makowski, T.; Piorkowska, E.; Rajkowska, K.; Kunicka-Styczyńska, A.; Biela, T. PLA/β-CD-based fibres loaded with quercetin as potential antibacterial dressing materials. *Colloid Surf. B-Biointerfaces* **2020**, *190*, 110949. [CrossRef]
43. Croitoru, A.-M.; Moroșan, A.; Tihăuan, B.; Oprea, O.; Motelică, L.; Trușcă, R.; Nicoară, A.I.; Popescu, R.-C.; Savu, D.; Mihăiescu, D.E.; et al. Novel Graphene Oxide/Quercetin and Graphene Oxide/Juglone Nanostructured Platforms as Effective Drug Delivery Systems with Biomedical Applications. *Nanomaterials* **2022**, *12*, 1943. [CrossRef] [PubMed]
44. Krishna Kumar, K.; Devendiran, M.; Senthil Kumar, P.; Sriman Narayanan, S. Quercetin-rGO based mercury-free electrode for the determination of toxic Cd (II) and Pb (II) ions using DPASV technique. *Environ. Res.* **2021**, *202*, 111707. [CrossRef] [PubMed]
45. Makowski, T.; Kowalczyk, D.; Fortuniak, W.; Jeziorska, D.; Brzezinski, S.; Tracz, A. Superhydrophobic properties of cotton woven fabrics with conducting 3D networks of multiwall carbon nanotubes, MWCNTs. *Cellulose* **2014**, *21*, 4659–4670. [CrossRef]
46. Razmara, R.S.; Daneshfar, A.; Sahraei, R. Solubility of Quercetin in Water + Methanol and Water + Ethanol from (292.8 to 333.8) K. *J. Chem. Eng. Data* **2010**, *55*, 3934–3936. [CrossRef]
47. *ISO 20645:2004*; Textile Fabrics—Determination of Antibacterial Activity—Agar Diffusion Plate Test. International Organization for Standardization: Geneva, Switzerland, 2004.
48. Smith, C.W.; Dole, M. Specific heat of synthetic high polymers. VII. Polyethylene terephthalate. *J. Polym. Sci.* **1956**, *20*, 37–56. [CrossRef]
49. Mehta, A.; Gaur, U.; Wunderlich, B. Equilibrium melting parameters of poly(ethylene terephthalate). *J. Polym. Sci. Polym. Phys. Ed.* **1978**, *16*, 289–296. [CrossRef]
50. Sibilia, S.; Bertocchi, F.; Chiodini, S.; Cristiano, F.; Ferrigno, L.; Giovinco, G.; Maffucci, A. Temperature-dependent electrical resistivity of macroscopic graphene nanoplatelet strips. *Nanotechnology* **2021**, *32*, 275701. [CrossRef]
51. Sharma, K.; Maiti, K.; Kim, N.H.; Hui, D.; Lee, J.H. Green synthesis of glucose-reduced graphene oxide supported Ag-Cu_2O nanocomposites for the enhanced visible-light photocatalytic activity. *Compos. B Eng.* **2018**, *138*, 35–44. [CrossRef]
52. Svyntkivska, M.; Makowski, T.; Shkyliuk, I.; Piorkowska, E. Electrically conductive crystalline polylactide nonwovens obtained by electrospinning and modification with multiwall carbon nanotubes. *Int. J. Biol. Macromol.* **2023**, *242*, 124730. [CrossRef]
53. Cassie, A.B.D.; Baxter, S. Wettability of porous surfaces. *Trans. Faraday Soc.* **1944**, *40*, 546–551. [CrossRef]
54. Mei, L.; Zhang, X.; Wang, Y.; Zhang, W.; Lu, Z.; Luo, Y.; Zhao, Y.; Li, C. Multivalent polymer—Au nanocomposites with cationic surfaces displaying enhanced antimicrobial activity. *Polym. Chem.* **2014**, *5*, 3038–3044. [CrossRef]
55. Chen, R.-B.; Zhang, K.; Zhang, H.; Gao, C.-Y.; Li, C.-L. Analysis of the antimicrobial mechanism of porcine beta defensin 2 against *E. coli* by electron microscopy and differentially expressed genes. *Sci. Rep.* **2018**, *8*, 14711. [CrossRef] [PubMed]
56. Shen, J.; Tian, Y.; Li, Y.; Ma, R.; Zhang, Q.; Zhang, J.; Fang, J. Bactericidal Effects against S. aureus and Physicochemical Properties of Plasma Activated Water stored at different temperatures. *Sci. Rep.* **2016**, *6*, 28505. [CrossRef]
57. da Costa, E.M.; Filho, J.M.B.; do Nascimento, T.G.; Macêdo, R.O. Thermal characterization of the quercetin and rutin flavonoids. *Thermochim. Acta* **2002**, *392–393*, 79–84. [CrossRef]

MDPI AG
Grosspeteranlage 5
4052 Basel
Switzerland
Tel.: +41 61 683 77 34

Materials Editorial Office
E-mail: materials@mdpi.com
www.mdpi.com/journal/materials

www.ingramcontent.com/pod-product-compliance
Lightning Source LLC
Chambersburg PA
CBHW061621210926
51850CB00009B/2390
* 9 7 8 3 7 2 5 8 7 5 2 4 5 *